Physics

for Paramedical and Allied Health Science Courses

Physics

for Paramedical and Allied Health Science Courses

G Purushothaman MSc, MPhil, PhD
Reader, Department of Physiology
Vivekanandha Dental College for Women and
Vivekanandha College of Allied Health Sciences
Elayampalayam, Tamil Nadu

S Qairunnisa MSc, PhD
Assistant Professor, Department of Physiology
Sri Venkateswara Medical College
Tirupati, Andhra Pradesh

CBSPD

CBS Publishers & Distributors Pvt Ltd

New Delhi • Bengaluru • Chennai • Kochi • Kolkata • Lucknow • Mumbai
Hyderabad • Jharkhand • Nagpur • Patna • Pune • Uttarakhand

Disclaimer

Science and technology are constantly changing fields. New research and experience broaden the scope of information and knowledge. The authors have tried their best in giving information available to them while preparing the material for this book. Although, all efforts have been made to ensure optimum accuracy of the material, yet it is quite possible some errors might have been left uncorrected. The publisher, the printer and the authors will not be held responsible for any inadvertent errors, omissions or inaccuracies.

Physics

for Paramedical and Allied Health Science Courses

ISBN: 978-93-5466-667-4

First Edition: 2023

Reprint: 2025

Published by Satish Kumar Jain and produced by Varun Jain for

CBS Publishers & Distributors Pvt Ltd

4819/XI Prahlad Street, 24 Ansari Road, Daryaganj, New Delhi 110 002, India
Ph: 011-23266838, 23289259 Website: www.cbspd.com
e-mail: delhi@cbspd.com

Corporate Office: 204 FIE, Industrial Area, Patparganj, Delhi 110 092, India
Ph: 011-49344934 Fax: 011-49344935 e-mail: publishing@cbspd.com; publicity@cbspd.com

Branches

- **Bengaluru:** Seema House 2975, 17th Cross, K.R. Road, Banasankari 2nd Stage, Bengaluru 560 070, Karnataka, India
  Ph: +91-80-26771678/79 Fax: +91-80-26771680 e-mail: bangalore@cbspd.com
- **Chennai:** 18/8B, Subbarayan Street, Shenoy Nagar, Chennai 600 030, Tamil Nadu, India
  Ph: +91-44-42032115, 26681266 e-mail: chennai@cbspd.com
- **Kochi:** 42/1325, 1326, Power House Road, Opposite KSEB, Power House, Ernakulum 682018, Kochi, Kerala, India
  Ph: +91-484-4059061–65 Fax: +91-484-4059065 e-mail: kochi@cbspd.com
- **Kolkata:** 147, Hind Ceramics Compound, 1st Floor, Nilgunj Road, Belghoria, Kolkata 700056, West Bengal, India
  Ph: +91-33-25330055/56 e-mail: kolkata@cbspd.com
- **Lucknow:** Basement, Khushuma Complex, 7 Meerabai Marg (behind Jawahar Bhawan), Lucknow 226001, UP, India
  Ph: +91-522-4000032 e-mail: tiwari.lucknow@cbspd.com
- **Mumbai:** PWD Shed, Gala No. 25/26, Ramchandra Bhatt Marg, Next JJ Hospital, Gate No. 2, Opp. Union Bank of India, Noorbaug Mumbai 400009, Maharashtra, India
  Ph: +91-22-66661880/89 e-mail: mumbai@cbspd.com

Representatives

- **Hyderabad** 0-9885175004
- **Jharkhand** 0-9811541605
- **Nagpur** 0-8692091830
- **Patna** 0-9334159340
- **Pune** 0-9664372571
- **Uttarakhand** 0-9716462459

Printed at Mudrak, Noida, UP, India

Forewords

It is a great pleasure for me to write the Foreword to the book *Physics for Paramedical and Allied Health Science Courses*, authored by Dr G Purushothaman and Dr S Qairunnisa. I know both of them for the past 10 years as a faculty at Medical College and they have wonderful experience in teaching human physiology, biophysics and health physics subjects. And they have published many research papers in the related subjects. The present book is covering the various aspects relating to concepts of physics to health in a systemic manner in easy language. And the book provides a comprehensive coverage to physical concepts in health sciences. Each chapter in the book is provided with a good illustration that helps the students to have thorough subject knowledge in a simplified manner in making the subject concepts easily understandable. I wish to congratulate both the authors for nice compilation of this wonderful book and I convey my heartfelt appreciation for the hardwork done by both the authors to bring out this wonderful textbook, which is greatly appreciated. I hope that the students will understand and enjoy the reading of this book and will find physics, much easier and interesting subject in health science.

Prof. KRS Sambasiva Rao
Vice-Chancellor
Mizorom University
(A Central University)
Aizawl, Mizoram

I wish to introduce the authors and pleasure to write the Foreword to the manual *Physics for Paramedical and Allied Health Science Courses* for undergraduate students. Though, the subject of physics in health science is an important notion for the allied health courses but not familiar among the allied health science students. The authors have clone outstanding job or compiling and editing this text into a valuable resource for each student. It is a comprehensive, illustrative, and well written text. The strength of the book is well-drawn illustrations, easy language, and contemporary concepts. I would like to congratulate the authors for bringing out this book successfully and I wish them all very best in their endeavour.

Prof. (Capt) Dr S Gokulanathan BSc MDS
Director, Dental and Paramedical Courses
Vivekanandha Educational Institutions
Tiruchengode, Tamil Nadu

A teacher's purpose is not to teach something but make students realize their inner potential and get oriented to establish a successful career. Physics is the field which is natural and teaches humanity to explore the usefulness of the natural phenomena for the betterment of life.

I am happy to introduce the authors and it gives me immense pleasure to introduce my student and an effective teacher, Dr G Purushothaman, with this book *Physics for Paramedical and Allied Health Science Courses*. It is a new textbook in the field filled with examples, illustrations, and explanations to facilitate students to appreciate the fundamental aspects of physics in health education.

I am sure this book fulfils the aim of quality education and to create equipped manpower. I wish authors for the successful reach of the book and to bring more quality teaching material.

Dr K Gunasekaran
Assistant Professor
CAS in Crystallography and Biophysics
University of Madras
Guindy Campus, Chennai

It gives me pleasure in introducing this book *Physics for Paramedical and Allied Health Science Courses*. I know the authors for the past 15 years. I had an opportunity to interact with them for research when I was in Madras Medical College, Chennai. The author, Dr G Purushothaman, is an accomplished undergraduate teacher in Allied Health Sciences and has mastered the art of interaction and communication for the needs of the students. Dr S Qairunnisa, the co-author, has added the component of physiological basis that is a connection between science to health that needed for the allied health students. I have gone through few chapters; the book elegantly meets the students' requirements. The principles, instrumentations, and experiments have been simplified with the help of illustration and with easy language.

I strongly recommend the book for the health sciences and paramedical students for the basic physics background in health sciences. I could observe the sincere efforts of the authors. The language of the book has been kept simple for the benefit of the beginners in their learning career.

I appreciate the pains and hardwork that has gone in developing the document and I sincerely wish the endeavours a grand success.

Dr G Sathya Narayanan
Department of Physiology
Dr. ALM PG Institute of Basic Medical Sciences
University of Madras
Taramani Campus, Chennai

Preface

I take immense pleasure in introducing the title *Physics for Paramedical and Allied Health Science Courses*. This book mainly concentrates on • basic physics, • basic electronics, • thermodynamics, • optical principles, and • physical principles of instruments in laboratories that are required by undergraduate students of allied health sciences. The content has been compiled according to the syllabuses of the courses in allied health sciences conducted by TN Dr MGR Medical University and other similar Indian universities.

An attempt has been made to provide a strong basis to imbibe the common physical, radiological and physiological concepts. Most of the courses in physiology are ruled by the physical and biophysical principles and hence the subject of allied health physics may be called the 'physical concepts in radiology and living process'.

The primary aim of this book is to meet the requirements of allied health students precisely from the examination point of view, providing knowledge of fundamentals in the field of physics in healthcare, and in imparting relevant information on the important applied aspects of various topics.

Though it is a difficult task to cover a vast area of physics dealing with various concepts in health, the utmost effort has been made to include various chapters on health physics in the present volume. Though there may have been inevitable errors and omissions in this volume, sincere suggestions for the improvement of the text and graphics may kindly be brought to my notice, which will be thankfully acknowledged and incorporated in the next edition.

I acknowledge gratefully my indebtedness to Dr S Qairunnisa for her sustained encouragement while preparing the book and taking a keen interest in the contribution of chapters, correction of proofs, preparation of index, content, and going also through the typescript.

I am especially grateful to my beloved students—Ms Aneesa Fathima, Ms Varsha and Ms Vishnu Priya—for their timely assistance.

I am thankful to the management and the editorial department of CBS Publishers & Distributors Pvt Ltd for all the help and support in bringing out this book to match the national and international standards.

G. Purushothaman

Contents

5. HEAT AND THERMODYNAMICS 88–103

6. REVIEW OF PHYSICS—INSTRUMENTATION 104–114

7. SOUND 115–119

Chapter 1

Basic Concepts

1.1 WHAT IS PHYSICS?

Physics is defined as the study of properties of matters and interactions between fundamental constituents of the universe. The concept of physics is well explained by the interface of matters with fundamental and general laws. For that, we need to know about how the matters are measured and compared. At the same time, the experiments are also designed to establish the units for those measurements and comparisons. One purpose of physics (in allied health) is to study and conduct those principles in biological subjects.

Why physics concepts are so important, for example, if scientist wants to develop clocks to compare the time interval or sequence of two successive events with extreme accuracy. You may wonder whether such accuracy is needed or worth the effort of scientist. Here is one example of the worth is: Without clocks of extreme accuracy, the global positioning system (GPS), that is, now vital to worldwide navigation would be useless.

Physics concept is accepted and understand better by experimental observations and quantitative measurements, like all other sciences. To find the limited number of fundamental laws that overrule the natural phenomena to the biological subjects and to use it to develop more theories that predict the results which are the main objectives of physics. The fundamental laws used in developing theories are expressed in the language of mathematics, the tool that provides a bridge between theory and experiment in health science.

Summary

Physics is the branch of science that studies matter, energy and how they interact. Simply, ***Physics is about the study of properties of matter in the universe***.

1.2 MEASURING THINGS

Learning measurable quantities are measured in physics by using different kinds of tools and instruments, among these notable quantities are length, time, mass, temperature, pressure and electric current.

Physical quantities are compared with a **standard** and its own unit. The unique name for the unit is also assigned to measure the quantity. For example, the quantity of length is measured in **meter (*m*)**. The standard corresponds to exactly 1.0 unit of the quantity. As we can measure, the standard for length, which corresponds to exactly 1.0 m, is the distance traveled by light in a vacuum in a **unit** time, i.e. a certain fraction of second. We can define a unit and its standard in any way we care to. However, the important thing is to do so in such a way that scientist around the world will agree that our definitions are both sensible and practical.

Once we have set up a standard—say, for length—we must work out procedures by which ***any length*** whatever it may be, for example, sensible things from the radius of an atom, the wheelbase of a car, distance between two places or even the distance to the sun, can be expressed in terms of the standard. Rulers are commonly used to measure the length which give us one such procedure for measuring length. However, many of our comparisons must be indirect. You cannot use a ruler for certain measures, for example, to measure the radius of an atom or the distance from the earth to the sun.

Likewise, there are so many physical quantities which are having a problem to organize them. Fortunately, they are not all independent; for example, speed is the ratio of an object that travel in a unit time, simply length to a time. That is what we should have an international agreement—a small number of physical quantities, such as length and time and assign standards to them alone. We then define all other physical quantities in terms of these base quantities and their standards (called base standards). Speed, for example, is defined in terms of the base quantities length and time and their base standards.

Base standards must be both invariable and accessible. If we define the length standard as the distance between one's forehead and their index finger on a stretched arm, we certainly have an accessible standard length—but it will, of course, vary from person to person. The demand for precision in science and engineering pushes us to aim first for invariability and its accessibility. Then we can apply a great effort to make duplicates of the base standards that are accessible to those who need them in a standard manner.

1.3 STANDARDS OF LENGTH, MASS AND TIME

In physics, the laws are stated in terms of basic quantities which are length (L), mass (M), and time (T). All other quantities are also expressed in terms of these three.

In 1960, International System of Units abbreviated as SI, for the practical system of units and placed a rule for prefixes, derived units and the former supplementary units, and other matters. The SI system also proposed a set of standards for measure length, mass, and other basic quantities. In this system, the units of length, mass, and time are the meter, kilogram, and second, respectively. Other SI standards established by the committee are those for temperature (the kelvin), electric current (the ampere), luminous intensity (the candela), and the amount of substance (the mole).

Length: Length is the measurable quantities which are used for identifying the size of an object or distance from one place to other. Typical measures of length are like "how long an object is" or simply the distance between two points. It is used for identifying the size of an object or distance from one place to another. The length of an object is its extended dimension, that is, its longest side.

As per decision taken by 17th CGPM, 1983, Resolution that ***Length (meter (m) was redefined as the distance travelled by light in vacuum during a time of 1/299 792 458 second***. The definition of the meter in force since 1960, based upon the transition between the levels $_{2}p^{10}$ and $_{5}d^{5}$ of the atom of krypton 86, is abrogated. In effect, this latest definition establishes that the speed of light in vacuum is precisely 299 792 458 m per second.

Mass: In physics, mass is a quantitative measure of inertia, fundamental properties of all matters and is the measure of ***the amount of matter in an object***. Mass is usually measured in grams (g) or kilograms (kg).

In other way, mass measures the quantity of matter regardless of both its location in the universe and if the gravitational force applied on it which is described as weight. An object's mass is constant in all circumstances unless an external gravitational force act on it; force acts on mass is depends on gravity. It may vary from place to place. For example, moon's gravity and earth's gravity

Your mass on the earth and the moon is equal but your weight on the moon varies about one-sixth of your weight on the earth. That is because of the gravitational force on the earth.

Rest mass: To be more specific about mass, we need to mention the concept of rest mass, also known as **intrinsic mass**. It is an object's rest mass, that is, constant in all circumstances. Albert Einstein showed that mass actually increases with speed, although the effect is only significant at speeds that are considerable fractions of the speed of light.

Mass conservation: Mass conservation is one of the fundamental laws in science. The product mass developed from the reactions is always to the mass of products that entered the reaction.

Time: Time is an ongoing sequence of events taking place, for example, the past, present, and future. The basic unit of time is the second. There are also minutes, hours, days, weeks, months, and years. We can measure time using clocks.

Time is defined as the duration between two events and it is also called time interval. In the international system of measurement (SI) the second (s) is the basic unit of time.

1.4 SI BASE UNITS AND THE RADIAN AND STERADIAN

The definitions of the SI supplementary units, the radian and steradian, which are now interpreted as SI derived units and generally accepted.

SI derived units are uniquely defined only in terms of SI base units.

Meter (m): The meter is the length of the path travelled by light in vacuum during a time interval of 1/299 792 458 of a second.

Kilogram (kg): The kilogram is the unit of mass; it is equal to the mass of the international prototype of the kilogram.

Second (sec): The second is the duration of 9 192 631 770 periods of the radiation corresponding to the transition between the two hyperfine levels of the ground state of the cesium-133 atom.

Ampere (amp): Often shortened to "amp", is the base unit of electric current in the International system of units (SI).

The ampere is that constant current that produce attractive force of 20×10^{-6} newton per meter of length between two straight, parallel conductors of infinite length and negligible circular cross section, and placed 1 meter apart in vacuum. A force would produce between these conductors are equal to 231027 newton per meter of length. This is also called **ampere** force law.

Kelvin (K): The kelvin, unit of thermodynamic temperature, is the fraction 1/273.16 of the thermodynamic temperature of the triple point of water which is a hundredth of a degree Celsius above the freezing point.

Mole (mol)

1. The mole is the amount of substance of a system which contains as many elementary entities as there are atoms in 0.012 kilogram of carbon 12. It is also defined as exactly 6.02214076 1023 particles, which may be atoms, molecules, ions, or electrons.
2. When the mole is used, the elementary entities must be specified and may be atoms, molecules, ions, electrons, other particles, or specified groups of such particles.

In the definition of mole, it is understood that unbound atoms of carbon 12, at rest and in their ground state, are referred to note that this definition specifies at the same time the nature of the quantity whose unit is the mole.

Candela (cd): The candela is the luminous intensity, in a given direction, of a source that emits monochromatic radiation of frequency 54031012 Hertz and that has a radiant intensity in that direction of (1/683) watt per steradian.

Radian (rad): The radian is the plane angle between two radii of a circle that cut off on the circumference an arc equal in length to the radius.

Steradian (sr): The steradian is the solid angle that, having its vertex in the centre of a sphere, cuts off an area of the surface of the sphere equal to that of a square with sides of length equal to the radius of the sphere.

1.5 TEMPERATURE AND HEAT

In physics, there is a big difference between temperature and heat. Although it is not at all obvious, temperature is related to the energy of the atoms and molecules of an object. Heat is the type of energy that flows from one body to another where there is a difference between their temperatures between the bodies.

People used to think heat was a fluid that flowed from hot objects to cold ones. But this is not true, as discovered by physicist **Benjamin Thompson** ***(1752–1814)***. It is important to realize the minute difference between heat and temperature scales of measuring hotness or coldness of any object. They are interconnected but do not mean the same thing ***(Heat ≠ Temperature)***, The core difference is that heat deals with thermal energy, whereas temperature is more concerned with molecular kinetic energy.

Heat describes the transfer of thermal energy between molecules within a system and is measured in Joules. Heat measures how energy moves or flows from one object. An object can gain heat or lose heat, but it cannot have in its own heat. Heat is a measure of change, never a property possessed by an object or system. Therefore, it is classified as a process variable.

Temperature describes the average kinetic energy of molecules within a material or system and is measured in *Celsius* **(°C)**, *Kelvin* **(K)**, *Fahrenheit* **(°F)**, *or Rankine* **(R)**. It is a measurable physical property of an object—also known as a state variable. Other measurable physical properties include velocity, mass, and density.

NOTE

In reality, coldness does not exist, simply coldness is the absence of heat, that is what we experienced cold, in the absence of heat. It is clashing atom of energy.)

1.6 THERMODYNAMICS

It is the branch of physics that deals with the study of transfer of heat from one object into other object in the forms of energy and vice-versa.

A thermodynamical system is said to be in thermally equilibrium when macroscopic variables (like pressure, volume, temperature, mass, composition, etc.) that characterize the system do not change with time.

1.6.1 Thermodynamical System

Thermodynamical system state is an assembly of an extremely large number of particles whose macroscopic variables can be expressed in terms of pressure, volume and temperature.

Thermodynamic system is classified into the following three systems (Fig. 1.1):

i. **Open system:** It exchanges both energy and matter with its surroundings.

ii. **Closed system:** It exchanges only energy (not matter) with its surroundings.

iii. **Isolated system:** It exchanges neither energy nor matter with its surroundings.

A thermodynamic system is not always in equilibrium. For example, a gas allowed to expand freely against vacuum. Similarly, a mixture of petrol vapor and air, when ignited by a spark is not an equilibrium state. Equilibrium is acquired eventually with time.

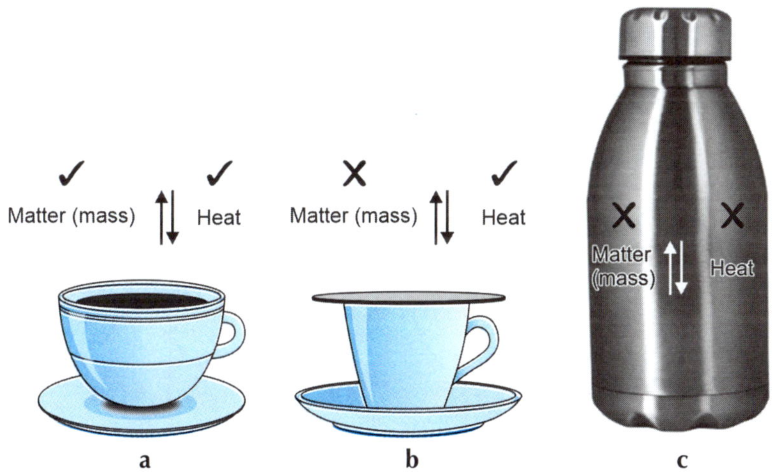

Fig. 1.1: Thermodynamical system: (a) Open system, (b) closed system and (c) isolate system

1.7 LAWS OF THERMODYNAMICS

The basic four laws of thermodynamics define fundamental characterize physical quantities such as temperature, energy, and entropy. These characteristics are in thermal equilibrium. The laws describe how these quantities behave under various circumstances, and forbid certain phenomena.

Four laws of thermodynamics are as follows.

1.7.1 Zeroth Law of Thermodynamics

It is also known as law of thermal equilibrium. It has given the name of zero because it was established after first law and second law. So, it was put forward to both the first and second as it provides logical basis for the concept of temperature of system. If two systems

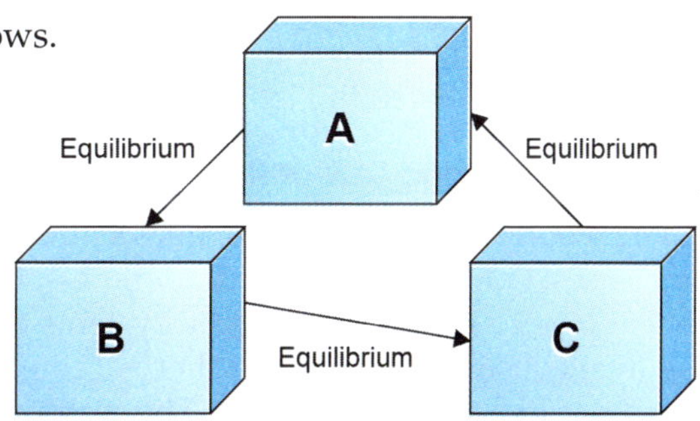

Fig. 1.2: Zeroth law of thermodynamics

are in thermal equilibrium with a third system, they are in thermal equilibrium with each other (Fig. 1.2). This law helps define the notion of temperature.

According to this law, if systems B and C separately are in thermal equilibrium with another system A, then systems B and C will also be in thermal equilibrium with each other. This is also the principle by which thermometers are used.

1.7.2 The First Law of Thermodynamics

The first law of thermodynamics, also known as law of conservation of energy, states that ***energy can neither be created nor to be destroyed; energy can only be transferred or changed from one form to another***. For example, turning on a light would seem to produce energy; however, it is electrical energy that is converted in that way energy is conserved.

A way of expressing the first law of thermodynamics is that any change in the internal energy (ΔE) of a system is given by the sum of the heat (q) that flows across its boundaries and the amount of work (w) is done on the system by the surroundings:

$$(\Delta E = q + w)$$

This law says that there are two kinds of processes that are heat and work, which can lead to a change in the internal energy of a system. Since both heat and work can be measured and quantified, this is the same as saying that any change in the energy of a system must result in a corresponding change in the energy of the surroundings that is outside of the system. In other words, energy can neither be created nor destroyed. If heat flows into a system or the surroundings to do work on it, the internal energy increases and the sign of q and w are positive. Conversely, heat flow out of the system or work done by the system (on the surroundings) will be at the expense of the internal energy, and q and w will therefore be negative.

Similarly, the internal energy E of system gas is increased by transferring heat q into the gas causing the gas molecules to move faster. If heat enters the gas, q will be a positive number. Conversely, we can decrease the internal energy of the gas by transferring heat out of the gas. We can achieve this by placing the container in an ice bath. If heat exits the gas, q will be negative number. This sign convention for heat q is represented in Fig. 1.3.

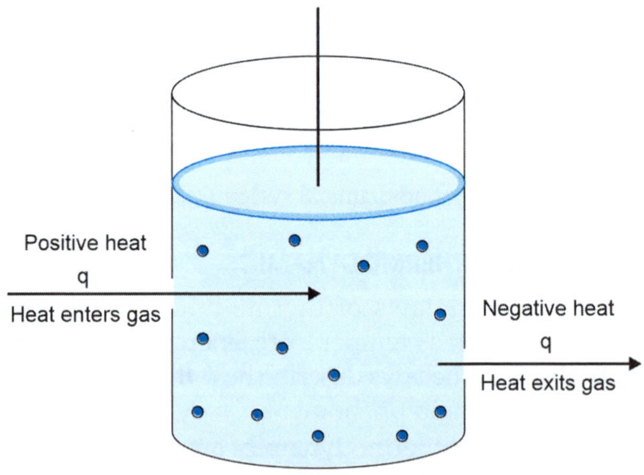

Fig. 1.3: The second law of thermodynamics

1.7.3 The Second Law of Thermodynamics

The second law of thermodynamics states that the entropy of any isolated system always increases. Entropy (degree of randomness, Fig. 1.4.) is a measure of the disorder and exists a useful variable in a system. All systems gain entropy at the end time. Isolated systems spontaneously evolve towards thermal equilibrium which is the state of maximum entropy of the system. More simply put: The entropy of the universe (the ultimate isolated system) only increases and never decreases.

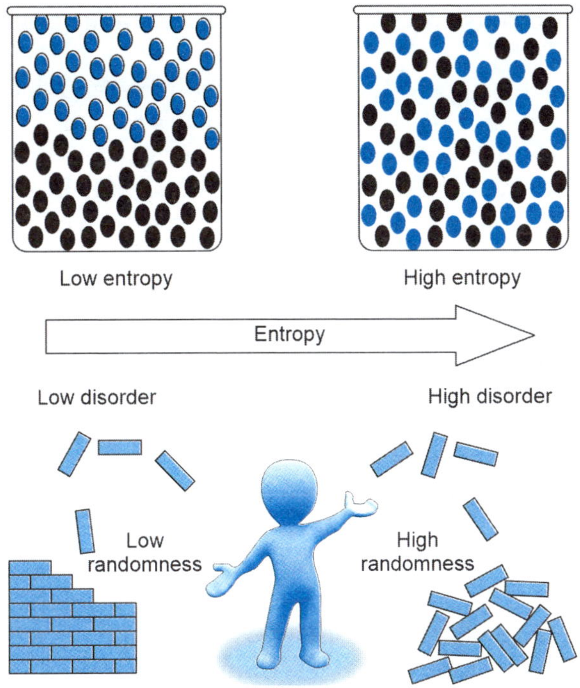

Fig. 1.4: The second law of thermodynamics

A simple way to think of the second law of thermodynamics is that a room, if not cleaned and tidied, will invariably become messier and more disorderly with time regardless of how careful one is to keep it clean. When the room is cleaned, its entropy decreases, but the effort to clean it has resulted in an increase in entropy outside the room that exceeds the entropy lost.

1.7.4 The Third Law of Thermodynamics

The third law of thermodynamics states that in a closed system the entropy (disorder) of a system approaches a constant value as the temperature approaches absolute zero (Fig.1.5). The entropy of a system at absolute zero is typically zero, and in all cases is determined only by the number of different ground states it has. Specifically, the entropy of a pure crystalline substance (perfect order) at absolute zero temperature is zero. This statement holds true if the perfect crystal has only one state with minimum energy.

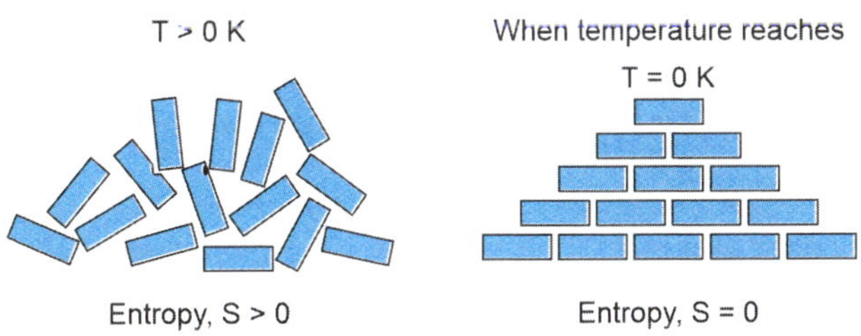

Fig. 1.5: The third law of thermodynamics

1.8 ACOUSTICS

The branch of physics that is concerned with the study of sound is known as acoustics. It also defined as the study of mechanical wave in solid, liquid and gas.

In general, ***Acoustics is the science that deals with the study of sound and its production, transmission, and effects.***

Acoustician is a person who studies about acoustics and ***acoustical engineering*** is the field where the person is working in the acoustics technology. The main application of acoustics is to make the music or speech sound as good as possible. It is achieved by reducing the sound barriers and increasing the factors that help in proper transmission of sound waves. Initially, acoustics was used only in industries which are based on sound like an auditorium, theatre but today, the application of acoustics has spread to many fields like medicine, warfare, architectural industries, etc.

1.8.1 Acoustic Energy

The commotion energy that passes through a matter in the form of wave is called acoustic energy. It is also concerning the mechanical vibration from its own components. Any acoustic event has the following stages.

- Cause or generating mechanism
- Acoustic wave propagation
- Reception effect

Propagation of sound: When a sound is produced, it travels like a wave in the form of vibration through air and when it reaches a person through ear, it is interpreted as sound. ***The process in which some form of sound is converted into sonic energy where producing a sound wave is called transduction process***. Sound waves carry its energy throughout the propagating medium. The acoustic wave equation is the fundamental equation that describes sound wave propagation. Wave propagation is the key process in any of the acoustic event. Sound propagates in liquids as a ***pressure wave*** and in solids as ***longitudinal*** or ***transverse waves***.

1.8.2 Types of Acoustics

1.8.2.1 Environmental Noise

In everyday life, noise that created in a roadways and vibrations are created in a big industries which are all concerned in environmental noise. In addition, railways, aircraft and general activities that also creates an unwanted or harmful noise. This noise which affects human based on their exposure, it may vary from emotional to physiological and psychological.

Environmental noise can also carry a sense of energy (lively) in a particular areas, which can be desirable. However, the adverse effects of noise exposure could include: interference with speech or other 'desired' sounds, annoyance, sleep disturbance, anxiety, hearing damage and stress-related emotional and cardiovascular health problems.

1.8.2.2 Musical Acoustics

Musical acoustics is concerned with the study of physics of music, i.e. how sounds are used to make music that helped in representing the pitch and duration of a sound in musical notation. There are so many areas of musical acoustics which include human voice, musical instruments, and music therapy.

The following three concepts can be considered the standard measuring units of music acoustics Hertz, cent and decibel.

Hertz is the standard measure of the frequency of oscillations in a wave motion. The frequency is 1 Hz when one oscillation occurs in one second. Similarly, when 1000 oscillations occurs in one second, then the frequency is 1000 Hz, or 1 kHz (kilohertz).

Frequency is nothing but number of cycles per second.

The unit most commonly used to measure micro-intervals is called **cent** (German Cent, Latin centum, meaning "one hundred"). It stands for one hundredth of an equal-tempered semitone. In other words, one octave consists of 1200 cents.

The **decibel (dB)** is used to measure the intensity level of a sound, but it is also widely used in electronics, signals, and communication industries. The dB is a logarithmic way of describing a ratio. The ratio may be power, sound pressure, voltage or intensity or several other things.

1.8.2.3 Ultrasounds

Ultrasounds are the sounds with a frequency greater than the human audible limit (20 Hz to 20 kHz, beyond the limit is called ***noise***). However, there is no difference in physical properties when compared to normal sound. Ultrasound is used in many fields. Ultrasonic devices are used in measuring distances and in detecting objects. Vast area of ultrasound physics are in medicine and imaging technology.

1.8.2.4 Infrasounds

Infrasounds are the sounds with a frequency of less than 20 Hz. The study of such sounds is called infrasonics. Applications include detection of petrol formation under the earth and the possibility of earthquakes.

Vibration and dynamics: It is the study of how mechanical systems vibrate and interact with their environment. Applications include vibration control which helps to protect a building from earthquakes and ground vibrations used in railways. Vibrational property may also define as the heaviness of an object.

1.9 POWER

Power is the rate at which the work is done or it is the amount of energy consumed per unit time. Having no direction, it is a scalar quantity. Power is increased if work is done faster or energy is transferred in less time. In the SI system, the unit of power is the joule per second (J/s), known as the watt (P = W/t).

1.10 ENERGY

Energy is defined as the capacity of a physical system to perform a work. Energy is a conserved quantity and the law of conservation of energy states that energy can neither be created nor to be destroyed but can only be converted from one form to another. The SI unit of energy is Joule.

Units of energy: The international system of units of measurement of energy is **Joule**. The unit of energy is named after James Prescott Joule. Joule is a derived unit and it is equal to the energy expended in applying a force of **one newton through a distance of one meter**.

1.10.1 Forms of Energy

Energy exists in several forms such as heat, kinetic or mechanical energy, light, potential energy, and electrical energy.

- **Heat:** Heat or thermal energy is energy from the movement of atoms or molecules. It may be considered as energy relating to temperature.
- **Kinetic energy:** Kinetic energy is energy of motion. A swinging pendulum has kinetic energy.
- **Potential energy:** This is energy due to an objects in position. For example, a ball kept on a table has potential energy with respect to the floor because gravity acts upon it.
- **Mechanical energy:** Mechanical energy is the sum of the kinetic and potential energy of a body. It is also defined as the energy of a substance or system by its motion.
- **Light:** Consists of photons, which are a form of energy produced when atom is heated.
- **Electrical energy:** This is energy from the movement of charged particles in an electric conductor such as protons, electrons, or ions.
- **Magnetic energy:** This form of energy results from a magnetic field.
- **Chemical energy:** Chemical energy is released or absorbed by chemical reactions. It is produced by breaking or making chemical bonds between atoms and molecules.
- **Nuclear energy:** This is energy from interactions with the protons and neutrons of an atom. Typically, this relates to the strong force. Examples are energy released by fission and fusion reactions.

1.10.2 Kinetic Energy and Work

Kinetic energy: Kinetic K is energy associated with the *state of motion* of an object. If the object moves faster, it has high kinetic energy. When the object is stationary, its kinetic energy is zero. For an object of mass m whose speed v is well below the speed of light,

$$K = \frac{1}{2}mv^2 \text{ (kinetic energy).}$$

Work: Work W is energy transferred to or from an object by means of a force acting on the object. Energy transferred to the object is positive work, and energy transferred from the object is negative work. The work done by force is the product of the force and the distance moved in the same direction (displacement).

$$\text{Work} = \text{force} \times \text{displacement}$$

1.11 FORCE AND DERIVATIVES

Force: Force is a quantitative description of an interaction of object that causes a change in an object's motion. An object may increase its speed, slow down, or change direction in response to a force acting on it. Objects are pushed or pulled by forces acting on them.

Contact force: Contact force is defined as the force exerted when two objects are in physical and direct contact with each other. Other forces, such as gravitation and electromagnetic forces, can exert themselves even across the empty vacuum of space.

Units of force: Force is a vector quantity; it has both direction and magnitude. It is a study of motion, acceleration which are changes in velocities. It can cause an object to change in velocity as well its accelerate. The SI unit for force is the ***newton* (N)**. One newton of force is equal to 1 $kg \cdot m/s^2$. Force is also represented by the symbol **F**.

Force is proportional to acceleration. In calculus terms, force is the derivative of momentum with respect to time.

Newtonian mechanics: It is the study of the relationship between a force and the acceleration, it was first denoted by Sir Isaac Newton (1643–1727) and it is called **Newtonian mechanics**.

Physicists now view newtonian mechanics as a special concept of comprehensive theories. Still, it is a very important special case because it applies to the motion of objects ranging in size from the very small (almost on the scale of atomic structure) to astronomical levels (galaxies and clusters of galaxies) of variations.

1.12 NEWTON'S LAWS OF MOTION

The concept of force was originally defined by Sir Isaac Newton, in his three laws of motion. He explained gravity as an attractive force between bodies that possessed mass.

1.12.1 First Law

In an inertial reference frame, an object either remains at the state of rest or continues to move at a constant velocity, unless an external force acted upon it.

In other words, the first law states two things:

1. A stationary object will only move if there is an unbalanced force acting on it.
2. A moving object will only change its speed or direction of motion, if there is an unbalanced force acting on it.
   - The first part states that a book on a table is resting on it, unless an unbalanced force act on it.
   - A ball rolling along a flat surface will slow down and eventually come to halt. But this is because of friction, which is also a type of force that slows down the ball. A ball rolling down a slope is also affected by friction, but the force of gravity that causes it to move is stronger. In a place without the forces of friction, air resistance and gravity (e.g. in outer space) a moving object would keep on moving in a straight line if there were no force to slow it down or change its direction.

1.12.2 Second Law

The second law states that the rate of change of momentum of a body, is directly proportional to the force applied on it and this change in momentum takes place in the direction of the applied force.

$$F = ma \quad ...(1.1)$$

For example, weight is a force that we feel on earth caused by the gravity. Weight is calculated as

$$W = mg \quad ...(1.2)$$

Where, m is the mass of the object and g is the local gravitational acceleration (not to be confused with G, the universal gravitational constant), roughly equal to 9.8 meters per second^2 (32 feet per second^2) on earth.

We can express Newton's second law in terms of momentum. The momentum p of a particle is defined as the product of its mass m and velocity v:

$$P = mv \quad ...(1.3)$$

Newton's second law states that the resultant force acting on a particle equals the time rate of change of momentum of the particle

$$F = \frac{dp}{dt} \quad ...(1.4)$$

For a particle of fixed mass (constant m), from eq (1.4)

$$F = \frac{dp}{dt} = \frac{d(mv)}{dt} = \frac{mdv}{dt} = ma$$

$$\vec{F} = m\vec{a} \qquad ...(1.5)$$

Equ (1.5), is the equation of Newton's second law in vector form.

1.12.3 Third Law

Newton's third law states that for every action, there is an equal and opposite reaction. As you sit on your table, your body acts on the table with one force, and the table reacts on your body with another force. This is an **example** of **Newton's third law** of motion.

When one body exerts a force on a second body, the second body simultaneously exerts a force equal in magnitude and opposite in direction on the first body.

According to Newton's third law, interacting objects experience two forces—**action** and **reaction**. The size of the forces acting on one object equals the size of the forces acting on the second object. Additionally, the direction of the force on one object is opposite to the direction of the force on the second object. Let us consider a man attempting to lift a heavy weight. If the man can pull up with a force, then the weights in turn will pull down with a force of same value.

1.13 FUNDAMENTAL FORCES

There are four fundamental forces that govern the interactions of physical systems. Scientists continue to pursue an unified theory of these forces.

- Gravitation is the force that acts between masses. All particles experience the force of gravity. While the graviton has been proposed as the particle mediating gravity, it has not yet been observed.
- The electromagnetic force acts between electrical charges and the mediating particle is the photon.
- The strong nuclear force holds the nucleus of the atom together, mediated by gluons acting on quarks, antiquarks, and the gluons themselves.
- The weak nuclear force is mediated by exchanging W and Z bosons, and is seen in beta decay of neutrons in the nucleus. At very high temperatures, the weak force and the electromagnetic force are indistinguishable.

1.14 MASS-ENERGY EQUIVALENCE

Derivation: Force is defined as rate of change of momentum, *i.e.*

$$F = \frac{d(mv)}{dt} \qquad ...(1.6)$$

According to the theory of relativity, both mass and velocity are variable. Therefore,

$$F = \frac{d(mv)}{dt} = m\frac{dv}{dt} + v\frac{dm}{dt} \qquad ...(1.7)$$

Let the force F *act on the body which* displace the body through a small distance dx. Then, the increase in the kinetic energy (dEk) of the body is equal to the work done (Fdx).

Hence, $$dE_k = F\,dx = m\frac{dv}{dt}dx + v\frac{dm}{dt}dx$$

or $$dE_k = mv\,dv + v^2\,dm \qquad ...(1.8)$$

According to the law of variation of mass with velocity

$$m = \frac{m_0}{\sqrt{1 - v^2 / c^2}} \quad ...(1.9)$$

Squaring both sides,

$$m^2 = \frac{m_0^2}{1 - (v^2 / c^2)}$$

or

$$m^2c^2 = m_0^2c^2 + m^2v^2$$

Differentiating,

$$c^2 2m\, dm = m^2 2v\, dv + v^2\, 2m\, dm$$

or

$$c^2\, dm = mv\, dv + v^2\, dm \quad ...(1.10)$$

From equations (1.8) and (1.10), $dE_k = c^2\, dm$ ...(1.11)

Thus, a change in KE (dEk) is directly proportional to a change in mass *dm of an object*.

When a body is at resting state, its velocity is zero, (KE = 0) and $m = m^0$. When its velocity is v, its mass becomes m. Therefore, integrating equation (6),

$$E_x = \int_0^{-k} dE_k = c^2 \int_{m_0}^{m} dm = c^2(m - m_0)$$

$$\therefore \quad E_x = mc^2 - m_0c^2$$

This is the relativistic formula for KE

When the body is at rest, the internal energy stored in the body is m_0c^2. m_0c^2 is called the rest mass energy. The total energy (E) of the body is the sum of K.E. (Ek) and rest mass energy (m_0c^2).

$$\therefore \quad E = E_k + m_0c^2 = (mc^2 - m_0c^2) + m_0c^2 = mc^2.$$

$$\therefore \quad E = mc^2$$

This is Einstein's mass-energy relation.

Chapter

2

Basic Electronics

2.1 ELECTRONICS

The branch of physics which deals with conduction of current through vacuum or gas or semiconductor.

In general, electronics deals with electronic devices and its utilization on various fields. When a current flow through a vacuum or gas or semiconductor device that enable them to function and can define its properties and characteristics. Based on the properties and its characteristics of electronic device has gained importance of applications in many industries.

The following are the important characteristic functions of an electronic devices:

a. **Rectification:** Rectification is the process of converting alternative current (AC) into direct current (DC). The devices which convert AC power into DC power (Fig. 2.1) with high efficiency is called rectifier. The rectified DC supply can be used for charging batteries to store that can be used in many appliance like televisions, computers, battery chargers, mobile phone chargers, etc.

Fig. 2.1: Rectifier

b. **Amplification:** Amplification is the process of strengthening the weak input signal. The device that achieves the amplification of signal is called amplifier (Fig.2.2). The amplifiers are widely used in variety of ways. For example, in case of radiofrequency an amplifier is used to amplify this weak signal so that it can be heard loudly. Similarly, amplifiers are also used in public domain, television, etc.

Fig. 2.2: Amplifier

c. **Control:** In domestic and industries purposes, electronic devices have wide applications in automatic control. For example, controlling motors speed, refrigerator voltage fluctuation, etc. It can be controlled automatically within the applications of electronic device.

d. **Generation:** Electronic devices can convert DC power into AC power of any frequency (Fig. 2.3). So, it works as an oscillator, main application of oscillator is generating high, low and stable frequency. It can be used for annealing and hardening.

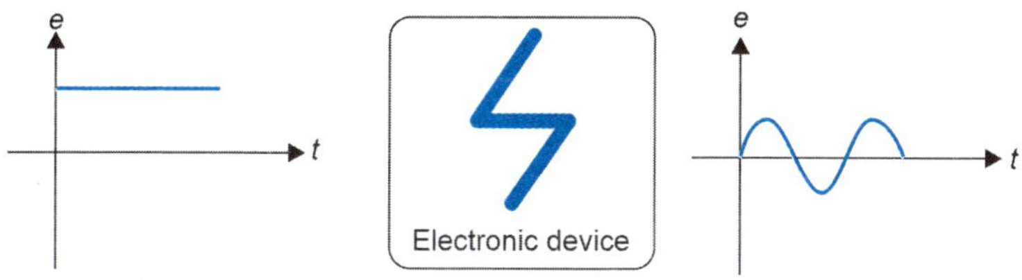

Fig. 2.3: DC to AC converter (Inverter)

e. **Conversion of light into electricity:** Electronic devices can convert directly light energy into electricity. This conversion of light into electricity is known as photo-electricity. Photo-electric devices are used in solar lights, sound recording, electronic industries, etc.

d. **Conversion of electricity into light:** Electronic devices can convert electricity into light. This characteristic can be used in television and radar.

2.2. ATOMIC STRUCTURE

Couple of centuries year back, many scientists attempted to explain the structure of the atom with the help of atomic models. Each of these models had their own merits and demerits and that was a key point to the development of modern atomic model. The most notable contributions to the field were by the scientists John Dalton, JJ Thomson, Ernest Rutherford and Niels Bohr.

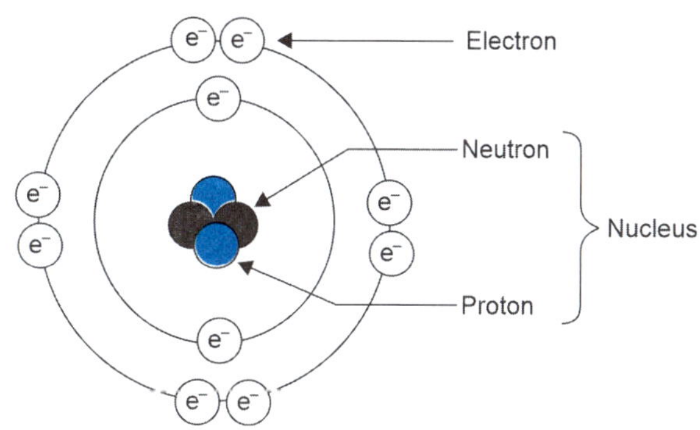

Fig. 2.4: The structure of atom

According to that, atomic structure refers to the structure of an atom comprising a nucleus at its centre in which the protons have positive charge and neutrons have neutral charge. The negatively charged particles is called electrons which revolves around the nucleus (Fig. 2.4).

Protons: Protons are positively charged subatomic particles. The charge of a proton is +1e, which corresponds to approximately 1.602×10^{-19}

The mass of a proton is approximately 1.672×10^{-24} g

The charge of a proton is equal and opposite to that of an electron. Protons are over 1800 times heavier than electrons.

The total number of protons or electrons in an atom is the atomic number of the element.

Neutrons: The mass of a neutron is almost the same as that of a proton, i.e. 1.674×10^{-24} g

Neutrons are electrically neutral particles and carries null charge.

Different isotopes of an element have the same number of protons but vary with the number of neutrons present in nuclei of an atom.

Atomic weight = No. of protons + No. of neutrons

Electrons: The charge of an electron is –1e, which approximates to -1.602×10^{-19} C

The mass of an electron is approximately 9.1×10^{-31} kg.

Due to the relatively negligible mass of electrons, they are ignored when calculating the mass of an atom.

As a whole, an atom is neutral. In an ordinary condition, the number of electrons is equal to the number of protons in an atom.

The electrons in an atom revolve around the nucleus in an orbits or paths. The following rules is determined the number and arrangement of electrons in the orbit.

1. The number of electrons present in the orbit is given by $2n^2$ where, n is the number of the orbit. For example,
   $n = 1$, first orbit, contains 2×1^2 = 2 electrons
   $n = 2$, second orbit, contains 2×2^2 = 8 electrons
   $n = 3$, third orbit, contains 2×3^2 = 18 electrons, etc.
2. The shells of an atom cannot accommodate more than 8 electrons, even if it has a capacity to accommodate more electrons. This is a very important rule called the Octet rule.
3. The stable state is either having an octet (8) or completely filled state (18) in the outermost shell.

2.3 STRUCTURE OF ELEMENTS

Atoms are made up of protons, neutrons and electrons. The atoms in the periodical tables are differed by its different arrangement and number of particles in the atoms. For example, the structure of carbon atom differs from the nitrogen atom and hence the two elements have different properties.

The atomic structure of any atom can be defined, if we know the atomic weight and atomic number of the element. Consider the atom of sodium

Atomic weight = 22.98
Atomic number = 11
No. of protons = No. of elecctrons = 11
No. of neutrons = 12

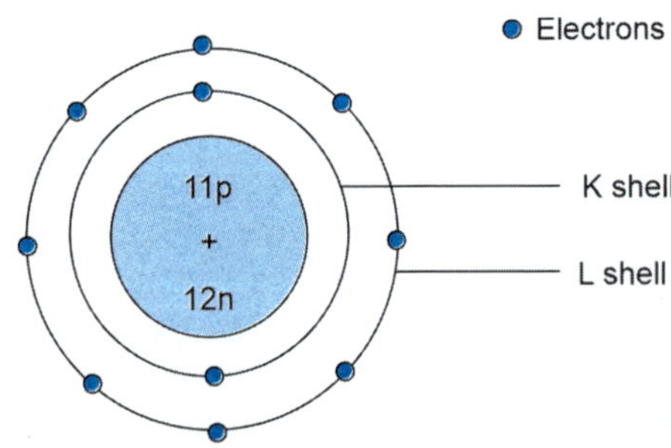

Fig. 2.5: Sodium

Sodium atom has 11 electrons which are arranged in different orbits as shown in Fig. 2.5.
The first orbit of sodium atom will have 2 electrons, the second 8 electrons. The atomic structure of all known elements, the electrons are arranged in $2n^2$ rule only.

2.4 ELECTRON

The electron is a subatomic particle, symbol. e^- or. β^-, whose electric charge is negative one elementary charge.

Electric charge: –1 e; $-1.602176634\ 10^{-19}$ C; ...

Composition: Elementary particle

Discovered: JJ Thomson (1897)
Mass: $9.1093837015(28) \times 10^{-31}$ kg; 5.485799 ...

2.5 ELECTRON ENERGY

Matters are in the three states with distinct physical forms that are solid, liquid, and gas. In gaseous substances, the arrangement of molecules is not so close. In liquids, the molecular arrangement is moderate. But, in solids, the molecules are so closely arranged, that the electrons in the atoms of molecules tend to move into the orbitals of neighboring atoms. Hence the electron orbitals overlap when the atoms come together.

Due to the overlapping of atoms in solids, instead of single energy levels, there will be bands of energy levels formed. These set of energy levels, which are closely packed are called **energy bands**.

When an electron moving around the nucleus of atom have two types of energies that are kinetic energy and potential energy. The total energy of the electron is the sum of these two energies. The electron revolves around the nucleus in different energy levels or in shells will vary with discrete distance from the nucleus. The lowest orbit electron has low energy level whereas the electron in the outermost orbit has higher energy. Hence, the energy levels increase as the distance from the nucleus increases.

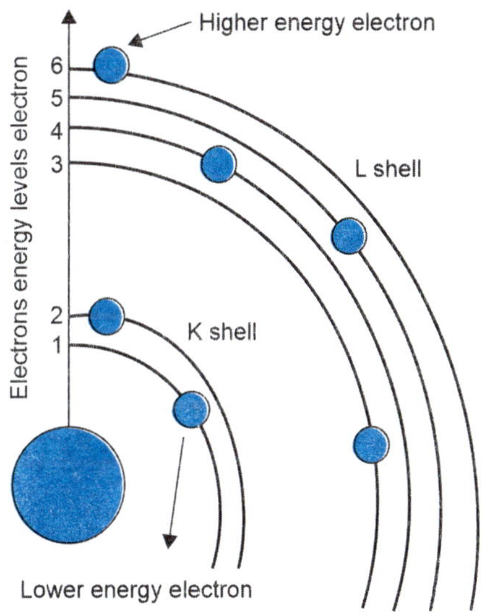

Fig. 2.6: Electron energy levels

There are many shells around the nucleus (Fig. 2.6) which are in the orbit as K, L, M, N and so on.

K shell: This shell ($n = 1$) is the closest to the nucleus. According to $2n^2$ rule, this shell can have only 2 electrons. For example, structure of helium with 2 electrons.

L shell: This shell ($n = 2$) is the second closest shell to the nucleus. According to $2n^2$ rule, this shell can have 8 stable electrons. For example, structure of neon is filled with 8 electrons.

M shell: This shell ($n = 3$) is the third closest shell to the nucleus. This shell can have 18 stable electrons. For example, structure of argon (inert gas) is filled with 18 electrons.

2.6 VALENCE ELECTRONS

The outer most orbit of an atom, the electrons are incomplete which is called valance electrons. This electrons in the valence orbit are less tightly bound to the atom. This is due to the force of attraction between the positively charged nucleus to the negatively charged electron decreases with increasing distance from the nucleus. Hence, the electrons in the higher energy levels are loosely bound to the nucleus of atom.

The valence electrons in the outermost orbit can determine the conducting properties of a material. On the basis of electrical conductivity, material can be classified into conductors, insulators and semiconductors.

i. If the material is a metal and a conductor, then the number of valence electrons of an atom is less than 4 (i.e. half of the maximum of eight electrons). For example, Na, Mg and Al are having 1, 2 and 3 valence electrons respectively (Fig. 2.7a to c).

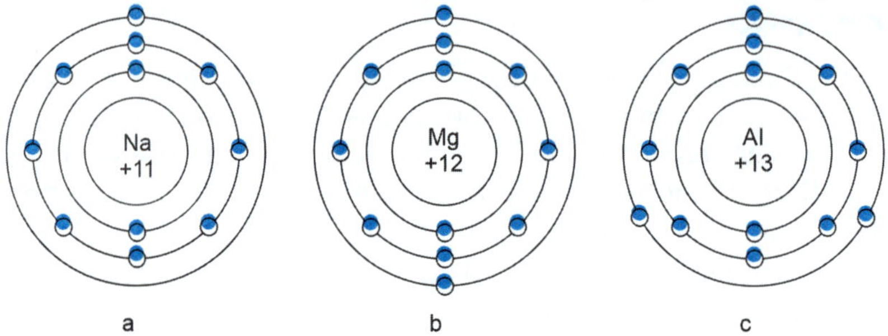

Fig. 2.7a to c: Electron configuration in metal and conductor

ii. If the material is a non-metal and an insulator, then the number of valence electrons of an atom is more than 4. For example, nitrogen, sulphur and neon which have 5, 6 and 8 valence electrons respectively (Fig. 2.8a to c).

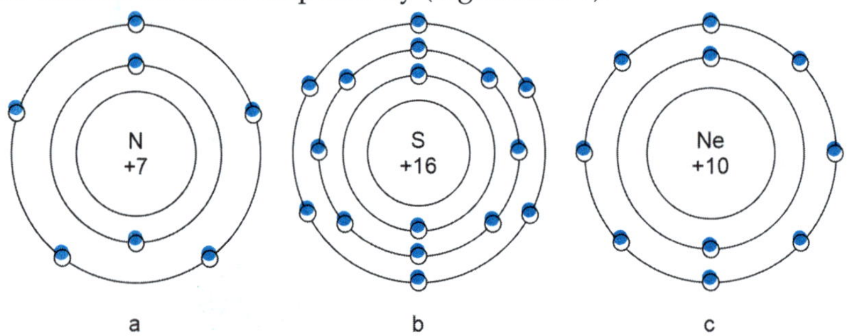

Fig. 2.8a to c: Electron configuration in non-metal and insulator

iii. If the material has both metal and non-metal properties and it is usually a semiconductor, then the number of valence electrons of an atom is 4 (i.e. exactly one-half of the maximum 8 electrons), For example, carbon, silicon and germanium has this properties (Fig. 2.9a to c).

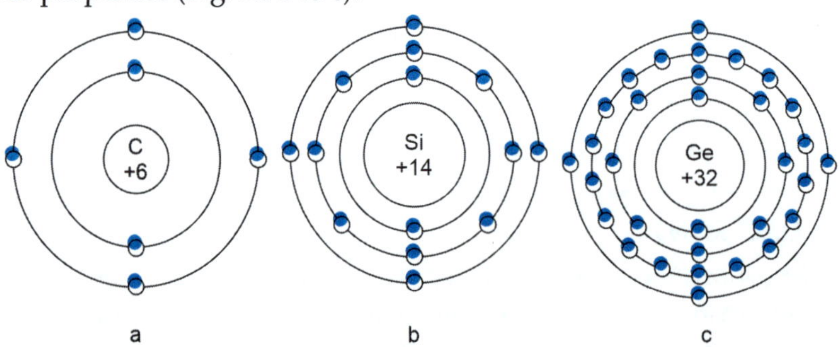

Fig. 2.9a to c: Electron configuration in semiconductor

2.7 FREE ELECTRONS

The electrons which are not attached to the nucleus of an atom, it is presented in the valence orbit and free to move when external energy is applied and it is called free electrons.

'he electrons in the valence orbit are called valence electrons which are loosely attached to ıe nucleus. A small external energy form heat or light is applied to the valence electrons ıen the electrons removed from the orbit and becomes free. The force of attraction of ıe nucleus does not influence on the free electron. The electrons flow in the material is alled an electric current. These moving electrons will transmit its energy to subsequent lectron and to next, then the material conduct currents. If the material does not have free lectrons, it does not conduct electric current. These electrons determine the conducting roperties of material.

On this basis of conducting property, material can be classified as, conductors, insulators nd semiconductors

i. A **conductor** is a substance in which a large number of free electrons conduct current. When potential difference is applied across a conductor, the free electrons move towards the positive terminal and creating electric current in the circuit.
ii. An **insulator** is a substance in which free electrons are not present at an ordinary temperature. Therefore, an insulator does not conduct current even under the influence of potential difference.
iii. A **semiconductor** is a substance in which very few free electrons are present at room temperature. Therefore, under the influence of potential difference, a semiconductor practically conducts no current.

.8 NUCLEAR FORCES

'he nuclear force is the force between nucleons (p^+ and n°) of an atom. It binds the nucleons ɔgether and the force can also exist between protons and protons, neutrons and protons r neutrons and neutrons. It can also hold the nucleons inside the nucleus. It acts as a lue to hold protons inside the nucleus. The charge of proton is +1e which tends to push way from each other with a strong electric field repulsive force, following Coulomb's aw (attraction or repulsion). But nuclear force is strong enough to hold them together nside the nucleus.

'roperties of Nuclear Force

It is attractive in nature but mainly repulsive. That is why the nucleus is held together without collapsing in itself.

Nuclear force is very short, even at 1 Fermi is also very small in distance between particles inside a nucleus. At this range, the nuclear force is much stronger than the repulsive Coulomb's force that pushes the particles away. However, if the distance is anything more than 2.5 Fermi, nuclear force is practically non-existent.

The nuclear force is identical for all nucleons. It does not matter about nucleons, once the Coulomb resistance is taken into consideration, nuclear force affects everything in the same way.

At a distance of less than 0.7 Fermi, this force becomes repulsive. It is one of the most interesting properties of nuclear force, as this repulsive component of the force is what decides the size of the nucleus. The nucleons come closer to each other till the point that the force allows, after which they cannot come any closer because of the repulsive property of the force.

.9 SHELL MODEL

n atomic and nuclear physics, the nuclear shell model is a model of the atomic nucleus which uses the Pauli exclusion principle *(in an atom or molecule, no two electrons can have the same set four electronic quantum numbers)* to describe the structure of the nucleus in terms of different energy levels. According to this model, the protons and neutrons are grouped inside the nucleus and it is also similar in number to extra-nuclear electrons in various shells outside the nucleus. The shells are considered as "filled" or fully occupied when they contain it has a specific number of protons or neutrons or both.

The shell model is also referred as the independent particle model because each nucleon that moves independently than other nucleons and is acted with an average nuclear field which is produced by the action of all the other nucleons.

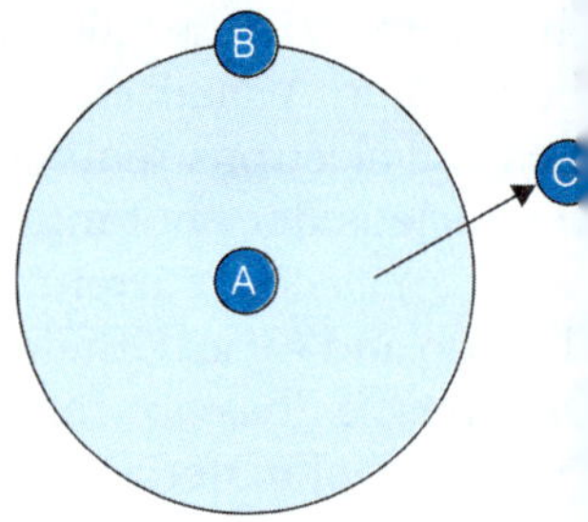

A—Nucleus, B—Electron, C—Extra-nuclear portion

Fig. 2.10: Shell model

2.10 INDICATION FOR SHELL MODEL

Stable nucleus should have a definite number of protons or neutrons. These numbers ar called *magic numbers*. The magic numbers are 2, 8, 20, 50, 82 and 126 which means the nucle containing 2, 8, 20, 50, 82 and 126 nucleons with sort of closed nuclear shell structures The following are the supporting interpretation of shell model.

i. Inert gases exhibit high chemical stability because of electrons are fully occupie and shells have no vacancy. Similarly, nucleons of the same kind electrons with it magic number also exhibit more than average stability.
ii. Isotope elements with magic number are having an abundance that greater tha 60%.
iii. The elements with magic number are more stable. For example Tin ($_{50}Sn$) has te stable isotopes and calcium ($_{20}Ca^{40}$) has six stable isotopes.
iv. The most stable isotope of radioactive series are uranium series, actinium series anc thorium series that decay to $_{82}Pb^{208}$ (Z = 82 and N = 126). Thus lead $_{82}Pb^{208}$ is the mos stable isotope with the numbers 82 and 126 indicate its stability.
v. Nucleus of specific atom having a number of neutrons which is equal to the magi number, cannot capture a neutron because the shells are closed and they cannc contain any extra neutron.
vi. Before the emission of β-decay some isotopes are excited above its nucleon bindin energy. For example, $_{8}O^{17}$, $_{36}Kr^{87}$ and $_{54}Xe^{137}$ for which N = 9, 51 and 83. It can b mentioned like 8 + 1, 50 +1, and 82 + 1 and can be taken as loosely bound neutror the neutron numbers 8, 50, 82 represent greater stability.

2.11 ATOMIC NUMBER AND MASS NUMBERS

Atoms of each and every element contain a certain characteristic number of proton Actually, the number of protons that determines the atomic number of an atom (e.g. a atoms with six protons are carbon atoms). In contrast, the number of neutrons for a give element can vary. Forms of the same atom that differ only in their number of neutron which are also called isotopes. Together, the number of protons and the number of neutron determine an element's mass number, therefore, mass number = protons + neutrons.

The atomic nucleus was first discovered by Rutherford in 1911. In his particle scatterin experiment, the atom consists of a very small nucleus *which is approximately* 10^{-14} *m i diameter* surrounded by orbiting electrons. As we know that, all atomic elementar particles in the nuclei are called protons and neutrons. A proton has a positive charge of the same magnitude as that of an electron but neutron is electrically neutral. Proton and neutron differed in different charge states of same atom of nuclei which are also called nucleon.

In an atom, the standard nuclear notation is $_{Z}X^{A}$, where, X–chemical symbol of the species, Z is the atomic number which indicates the number of protons and A is the mass number which indicates the total number of protons and neutrons of atom. N = Number of neutrons = A–Z.

For example, the chlorine nucleus $_{17}Cl^{35}$ has Z = 17 protons, A = 35 and N = 35–17 = 18 neutrons.

2.12 CLASSIFICATION OF NUCLEI

Atoms of different elements can be classified as follows:

i. **Isotopes** of an elements are nuclei of an atom with *the same atomic number* (Z) *but different in mass numbers* (A). For example, silicon (Si) isotopes of nuclei are $_{14}Si^{28}$, $_{14}Si^{29}$, $_{14}Si^{30}$, and $_{14}Si^{32}$. The isotopes of an element of all silicon atom contain the same number of protons but have different number of neutrons. The characteristic properties of isotopes are mainly depending on the nuclear charge of an atom. These isotopes are having same chemical behavior and whereas it is differing only in its physical mass.

ii. **Isobars**, elements are nuclei of atom *with the same mass number* (A), *but different atomic number* (Z). For example, oxygen and nitrogen are example of isobars atom $_{8}O^{16}$ and $_{7}N^{16}$. The isobars of an atoms are different elements and have different physical and chemical properties.

iii. **Isotones**, elements are nuclei of atom with an equal number of neutrons, that is, with the same N. For example, some isotones are $_{6}C^{14}$, $_{7}N^{15}$ and $_{8}O^{16}$, where, N = Number of neutrons = 8 in each atom.

iv. **Isomers**, there are some atoms with same Z and same A, but differ from one another only with its nuclear energy states and it is also exhibit differences in their internal structure. These nuclei can be distinguished by their different life times.

v. **Mirror nuclei**, nuclei with same mass number A, but differ with the proton and neutron numbers that are interchanged, which mean the number of protons in one in an atom and it is equal to the number of neutrons in the other atom. So that these two atoms are called mirror nuclei. For example, $_{4}Be^{7}$ atomic number, Z = 4 and number of neutrons, N = 3 and $_{3}Li^{7}$ atomic number, Z = 3 and number of neutrons = 4.

2.13 BOHR'S ATOMIC MODEL

Detailed explanation of atomic structure was given by Neils Bohr who is a Danish Physicist in 1913. He proposed the model which is modified form of Rutherford's model of an atom. Rutherford in his model, introduced the atom with positively charged nucleus which is surrounded by negatively charged electrons.

Bohr explained his modified atomic structure model with electrons that move in fixed orbit or shell and not somewhere in between and he also mentioned that each orbit has a fixed energy level. In his modified model, the electrons are occupied in shell based on their energy levels. He also found that electrons located away from nucleus has more energy, whereas the electrons close to the nucleus should have less energy.

Postulates of Bohr's atom model

1. In an atom, protons have positive charge that present inside the nucleus around which negatively charged electrons are revolving around the nucleus in circular orbits.
2. In certain permitted orbit the electrons can revolve only around the nucleus.
3. The electrons with certain fixed amount of energy can revolve around the nucleus in the permitted orbit. That means, the electrons with greater energy will revolve in the larger orbit (i.e. larger radius).
4. If an electron is stimulated with an additional energy like heat, light, etc. it is lifted the electron to the higher energy level. The atom is said to be in excited state. This state does not last longer, because the electron returns back to the original state. As it falls it gives back its energy in the form of heat, light or other radiations.

Figure 2.11 shows the orbital numbers positions of an atom and permitted energy levels. r_1, r_2, r_3 and r_4 are orbital radii of electrons. Consequently, radii between r_1 and r_2 or between r_2 and r_3 or between r_3 and r_4 are forbidden. Each orbit of an atom associated with fixed amount of energy. If an electron in the lower orbit is shifted to next orbit only when the energy should be equal to the higher orbital electron energy. So, it is excited in that state. When this electron jumps from higher orbit to its lower orbit, it will give back the acquired energy in the form of type of radiations.

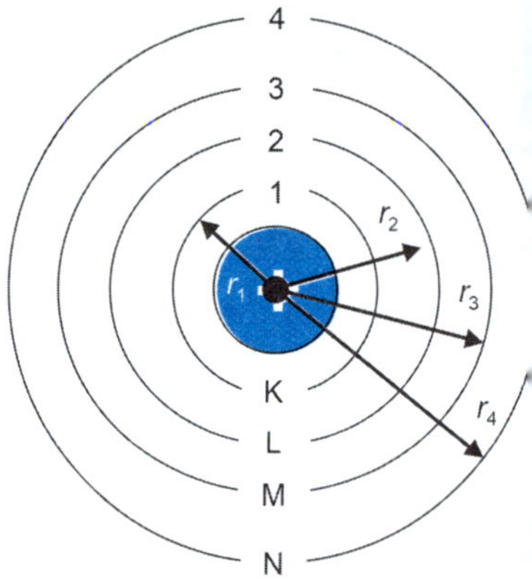

Fig. 2.11: Bohr's atom model

Limitations

- According to Zeeman's effect, the splitting of a spectral line into several components in the presence of a static magnetic field. But Bohr's atom model failed to explain the Zeeman effect on the spectra of atoms.
- In addition, strak effect also explains the splitting of a spectral line into several components in the presence of an electric field. It also failed to explain the stark effect on the spectra of atoms.
- According to Heisenberg uncertainty principle, the position and the velocity of an object both cannot be measured simultaneously. So it also violates the Heisenberg uncertainty principle.
- In case of larger atoms, it could not explain the spectra.

2.14 ENERGY LEVELS

According to Bohr's atom model, the orbit has fixed amount of energy with permitted level in the nucleus. Electron has greater energy when it revolves in the larger orbit. From the Bohr postulates, it is cleared that the electrons possess more energy than the inner orbital electrons.

To understand in a better way of electrons energy in different orbits is shown in Fig. 2.12. This is also called ***energy level diagram*** of an atom. The first orbit

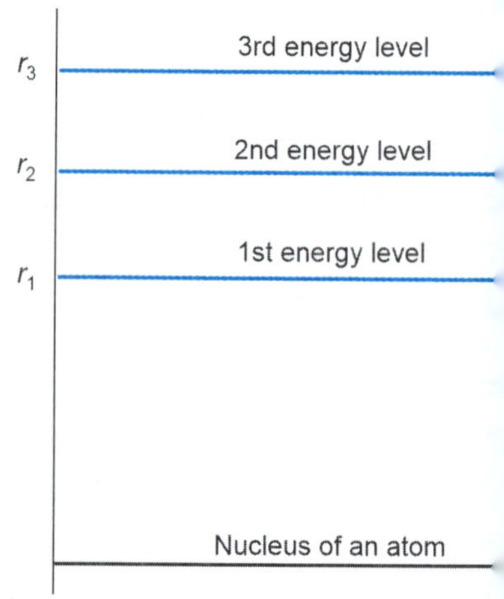

Fig. 2.12: Energy level diagram

indicates the 1st energy level, similarly second orbit indicates the 2nd energy level and so on. According the Bohr, the electron in the larger orbit should have greater energy and also higher in the energy level.

2.15 ENERGY BANDS

Energy bands of an atoms can be explained with its orbital levels. For example, in a single isolated atom, the electrons present in any orbit will have a definite energy. If an atom is a solid material which have great influenced by its neighboring atoms that are close to it. All atoms together, resulted combined electrons of all atom can have similar range of energies and it is considered to be a band instead of single energy. So, it is called ***energy band***.

With the help of following, Figure 2.13a shows the concept of energy can be understood easily with orbit radii r_1, r_2 and r_3. Figure 2.13b shows the energy levels of single isolated atom. Orbit of an atom has a discrete energy, considered the atom is in solid, the electron in any orbit can have a range of energies. For example, the electrons in the first orbit have discrete energies and it can occupy only two electrons as per $2n^2$ rule. All electrons in the first orbit will be exactly the same charge energy that will form a band called 1st energy band (Fig. 2.13c). Similarly, 2nd and 3rd energy bands also formed.

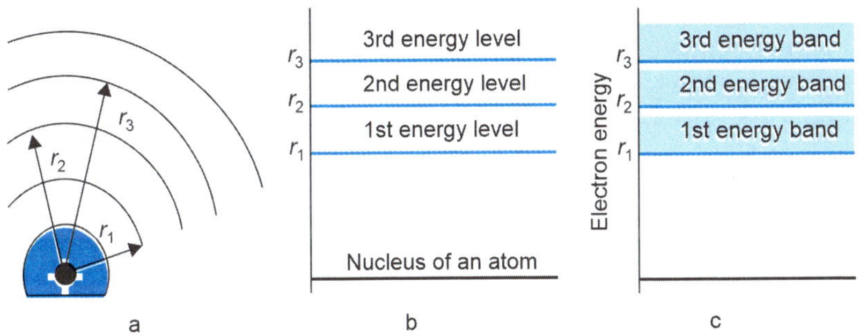

Fig. 2.13a to c: Energy levels in orbit, isolated atom and solid

2.16 ENERGY BANDS IN SOLIDS

In an isolated atom, energy bands of solid with K, L, M shells will be discussed below. The energy levels are depicted in Fig. 2.14 as corresponding bands of atom of solid. Depends on the atoms, number of energy bands are available in solids. The valance electrons are present in the outer most orbit is called valance band. It is more important range of energies (*i.e. band*) possessed by valence electrons is known as valence band.

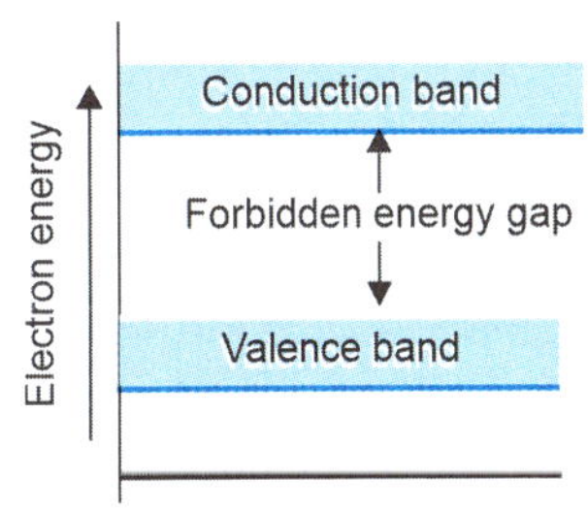

Fig. 2.14: Energy bands in solids

In an atom, valence band electrons are having highest energy. Depending upon atom, band may be completely filled or partially filled. For example, in case of inert gases, the valence band is fully occupied with electrons whereas the other materials are not fully filled. This partially filled valence band atom can accommodate still more electrons which can create a bond between other atoms.

In some type of materials like metals, the valence electrons are partially filled but loosely attached to the nucleus. Even under room temperature, some of these valence electrons may be removed from valence orbit and become free electrons which is more responsible for the conduction of current in the conductor. That is why these free electrons called conduction electrons. The free electrons are forming a range of energies (*i.e. band*) is called conduction band.

All electrons in the conduction band are called free electrons. If the material does not have such type of free electrons that will not conduct current. This type of material is called ***insulators***.

When a material acquired an external energy, the valence electrons move to the conduction band to conduct current. There is a gap between conduction band and valence band on the energy level diagram is called ***forbidden energy gap***.

The size of the forbidden energy gap can be determined by the valence electrons of an atom. If the valence electrons are bound more tightly to the atom, then the forbidden energy gap is more. In order to push the electrons from valence band to the conduction band, the electron should acquire an external energy which is greater than or equal to forbidden energy gap.

2.17 CLASSIFICATION OF SOLIDS WITH ENERGY BANDS

From the above discussion, the solid material of an atom with more free electrons are called good conductor of electricity whereas other material without free electrons are called insulators. There is also an intermediate class of material will act based on different condition is called semiconductor. The behavior of solid conductor and difference in electrical conductivity can be explained with the help of energy bands neatly. The electrons in the lower energy band are bound tightly with nucleus and it plays an important role in the conduction of electricity. The electrical behavior of material can be enlightened with the help of forbidden energy gap, valence bands and conduction bands.

Depending upon the electrical conducting property, the material can be classified as,

i. Conductors,

ii. Semiconductors, and

iii. Insulators.

2.17.1 Conductors

Electricity conducting material like copper, aluminium is substance which can allow the current to pass on material easily. This is mainly due to presence of free electrons. In terms of forbidden gap and energy band, the valence and conduction bands overlap each other with no forbidden energy gap as shown in Fig. 2.15. When this type of materials is placed between poles of a battery, due to no forbidden gap and overlapping of bands even at room temperature electrons will flow and conducts electric current in the material.

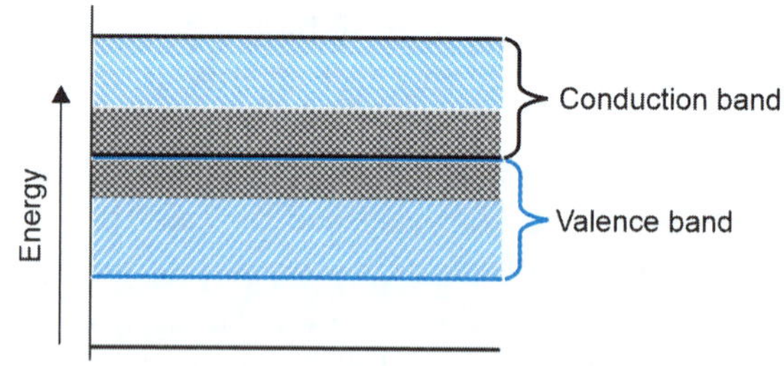

Fig. 2.15: Energy band diagram for conductors

2.17.2 Semiconductors

Material like silicon, germanium is substances whose conducting property lies between conductors and insulators. In terms of forbidden gap and energy band, the valence band is almost filled and conduction band is almost empty. In addition, the energy gap between valence and conduction bands is very small as shown in Fig. 2.16. A small electric field is more enough to push the electrons from the valence band to the conduction band. The forbidden energy gap is ≈1eV, i.e. in between valence and conduction bands. At very low temperature, the valence band is filled with electrons and the conduction band is empty. Therefore, a semiconductor behaves like an insulator at low temperature whereas at room temperature some electrons are crossing the gap and reach to the conduction band that imparting current. As the temperature of semiconductor material is increased, more electrons cross the gap and imparting more current. This shows that conductivity of material increased with increasing temperature.

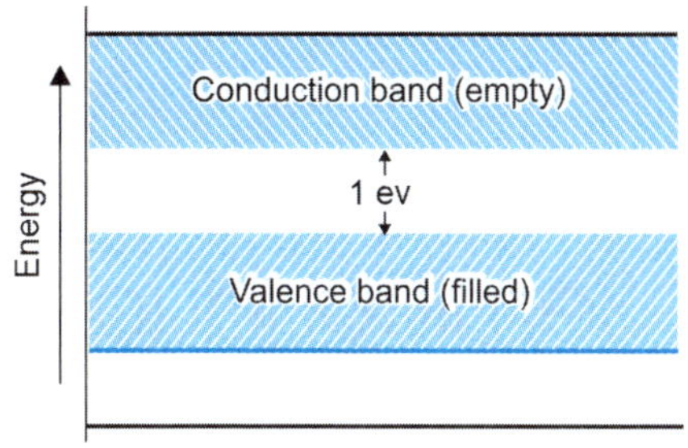

Fig. 2.16: Energy band diagram for semiconductors

2.17.3 Insulators

Electricity non-conducting material like wood, glass is substance which do not allow the current to pass on material. In terms of forbidden gap and energy band, the valence band is full whereas the conduction band is empty. In addition, the energy gap between valence and conduction bands is very large (≈15 eV) as shown in Fig. 2.17. Therefore, a very high electric field is require to push the electrons to the conduction band from valence band. Due the larger gap between band is reason for very low electrical conductivity in insulator at ordinary room temperature.

When the temperature is raised, very few numbers of electrons from valence band can shift to conduction band. Hence, resistivity of insulator decreases with increase in temperature.

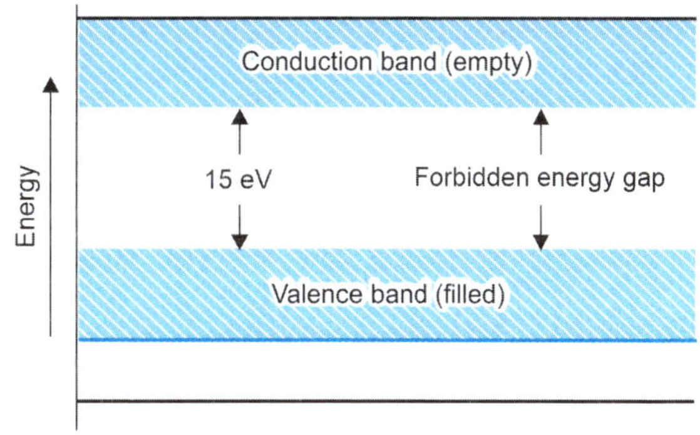

Fig. 2.17: Energy band diagram for insulators

Chapter

3

Semiconductor Physics

3.1 INTRODUCTION

In early 19th century, study of semiconductor materials are started. Substances like germanium (Ge), carbon (C), silicon (Si), tin (Sn), selenium (Se) are composed single species of semiconductor. There are numerous compound semiconductors, which are composed of two or more elements. Gallium arsenide (GaAs), mercury indium telluride ($HgIn_2Te_4$), aluminum gallium arsenide ($Al_xGa_{1-x}As$), etc. are the few examples of compound semiconductor.

Semiconductors have a numerous application in the electronic field and have some useful properties. Formerly, bipolar transistor plays a major role in this field before the invention of semiconductor which is used as rectifiers and photodiodes. In this chapter, Semiconductors will be discussed in the different aspects.

3.1.1 Semiconductor

As we discussed about semiconductor in the last chapter, it is type of materials which are having conducting property between conductors (metals) and non-conductors (insulators: Plastics). In physics, it is defined as the substance which has resistivity 10^{-4} ohm-cm to 10^3 ohm-cm and will be explained by the theories, properties, and mathematical method.

Properties of Semiconductors

i. The resistivity of semiconductor is present in between conductor and insulator.

ii. Semiconductors have negative temperature co-efficient of resistance which means the resistance on semiconductor will decreases with increase in temperature and vice-versa.

iii. By adding impurity (doping) to the pure form of semiconductor, substantial changes of conducting property are observed.

3.2 BONDING IN SEMICONDUCTORS

The semiconductor atoms of all elements are having electrons which are located on the outermost orbit that are incomplete. These atoms are detained together by the bonding of valence electrons. This is due to atoms are having tendency to have 4 electrons in the outer most orbit and it is ready to make a bond with another atom. These bonds are formed by sharing of valence electrons which are also called co-valent bonds. Figure 3.1 shows the covalent bonds among silicon atoms. A silicon atom has 4 valence electrons in the outer most orbit (valence orbit). Each individual atom bonded together in a regular and periodic manner. So, each atom is surround by 8 electrons. It consists of proton (positive charge) and neutron (neutral charge) surrounded by electrons. The electrons surrounding each atom are part of covalent bond. Therefore, each atom and 4 surrounding atoms, 8 electrons are being shared.

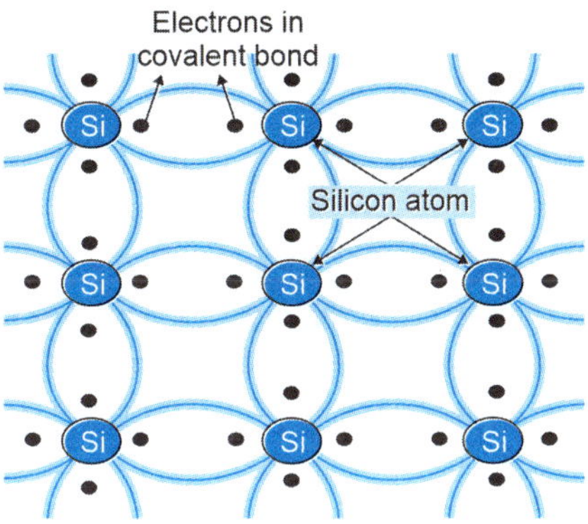

Fig. 3.1a: Covalent bond in Si

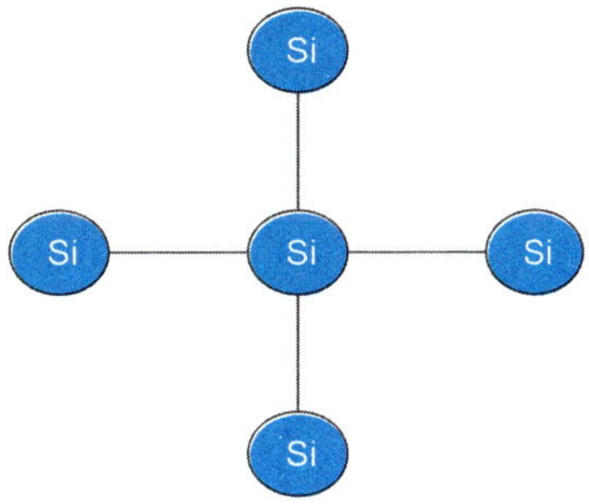

Fig. 3.1b: Bonding of Si

3.3 ENERGY BAND DIAGRAM OF SEMICONDUCTORS

As we know that a semiconductor is a material whose conductivity and resistivity lies between conductor and insulators. Conductivity of semiconductor is in the range of 10^{-6} to 10^4 ohm^{-1} cm^{-1} and resistivity is in the order of 10^{-4} ohm-cm (*doped*, refer to Subsection 3.6.2) to 10^3 ohm-cm (*undoped*, refer to section 3.6.1). The concept of energy band can be understood easily in terms of energy band description comprehensively.

Figure 3.2 shows the energy band diagrams of silicon and germanium respectively. It is observed that forbidden energy gap is very small in both silicon (1.1 eV) and germanium (0.7 eV). Therefore, minimum amount of energy which should be more than forbidden energy gap is enough for these valence electrons to jump the gap to reach conduction band. Even at room temperature with energy, it can reach the conduction band to conduct current. If the energy is provided, it cannot cross the gap and became an insulator. For this reason, it is called semiconductors.

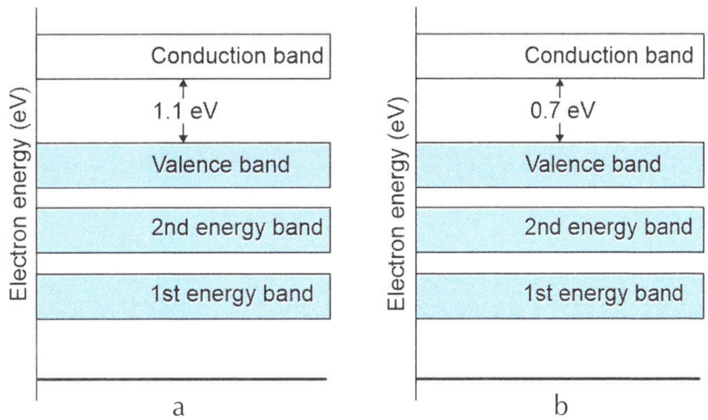

Fig. 3.2: Band diagram of (a) silicon and (b) germanium

3.4 EFFECT OF TEMPERATURE ON SEMICONDUCTORS

The conductivity of a semiconductor changes significantly with varying temperature. The conducting property of semiconductor with the effect of temperature as follows:

(i) At absolute zero: At absolute zero temperature (lowest temperature thermodynamic system, 0°K/–273.15°C/–459.67°F), electrons are tightly detained by the semiconductor atoms. The inner shell electrons are packed very tightly whereas the electrons in the valence shell are in covalent bond with other atoms. At this absolute temperature, covalent bonds are very strong enough to hold the electrons. So, at this condition it acts as an insulator as shown in Fig. 3.3.

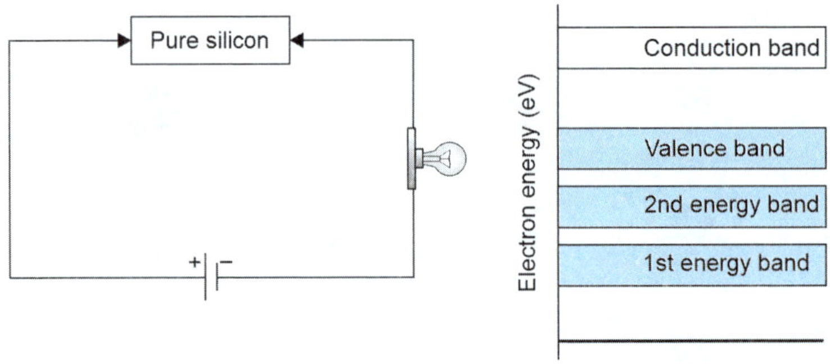

Fig. 3.3: Pure semiconductor conductivity as perfect insulator

On the basis of energy band description, electrons in the valence orbit is filled and there is large energy gap between valence and conduction band. Therefore, valence electron cannot reach the conduction band to conduct current. At absolute zero temperature, there is no free electrons and behaves as an insulator.

ii. **More than zero:** When increasing the temperature of material, covalent bonds of atom break due to the external thermal energy. The breaking of bonds creates free electrons. Finally, a few of free electrons are available to conduct current. This creates a low electric current in the circuit which is applied a supplied with a battery voltage V as shown in Fig 3.4.

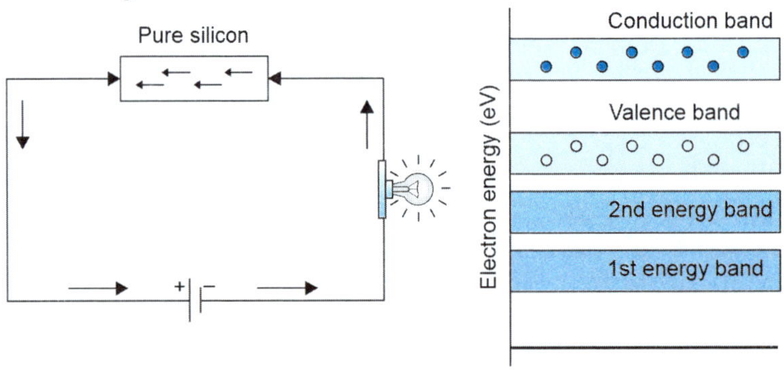

Fig. 3.4: Pure semiconductor conductivity when increase in temperature

From the above diagram, when the temperature of the material is increased that causes the valence electron received sufficient energy which shifts it to the conduction band to conduct current. When the electron is shifted to conduction band, a respective hole has been created in the valence band. It is also observed that hole also contribute for conduction of current.

3.5 CONTRIBUTION OF HOLE FOR CONDUCTION OF CURRENT

At normal room temperature, some of the covalent bonds in pure semiconductor break, that create free electrons. Due to impact of electric field, these free electrons contribute electric current in the circuit. When the temperature increases that produces thermal energy inside the material which removes electrons. In general, negative charge electron created from breaking of bond that leaves positive charge (hole) in the atom. For example, due to thermal energy in the material that removes one electron which causes a vacancy in the valence band. For one electron set free, one hole is created. Now we can understand the concept conduction of current in the circuit by holes is as follows.

In semiconductor material, the created hole shows a missing electron. Let us consider the valence electron position at P and it has become free electron due to thermal energy. This creates a hole in the co-valent bond position at P. The hole has a strong tendency to attract electron form neighboring atom which is in the position at Q. Now, the hole at P is filled by the electron Q. This results in the creation of hole at Q. Similarly, the valence electron at R may loss its bond to fill the hole at Q, thus, creating a hole at R. Thus, the hole having a positive charge has moved from P to R, i.e. towards the negative terminal of battery. This shows hole current in the circuit as shown in Fig. 3.5.

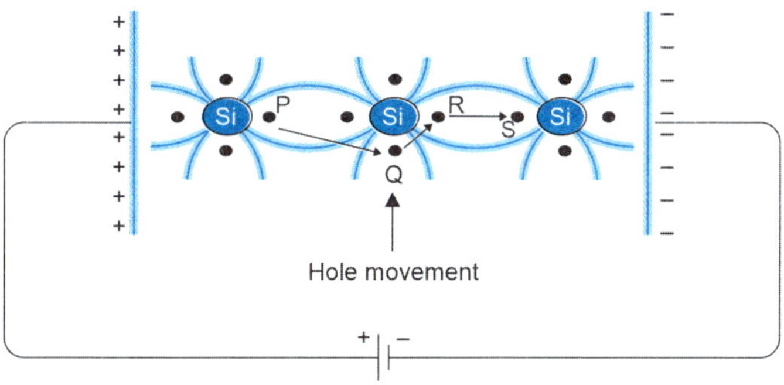

Fig. 3.5: Hole current in the pure form of semiconductor hole current

It is also noted that unlike normal current which is caused by flow of electrons the circuit hole current is due to the movement of valence electrons from once covalent bond to another bond. Obviously, the flow of current is against the direction of electron flow and it is in the direction of hole movement.

Let we understand this concept through an energy band diagram as shown in Fig. 3.6. Consider, crystal structure form of Si into energy band description, where PQRS path is the movement of hole which is like the direction of current whereas electron travel in SRQP direction. From this explanatory diagram, we can understand the concept of shifting of valence electrons to conduction band via forbidden energy gap.

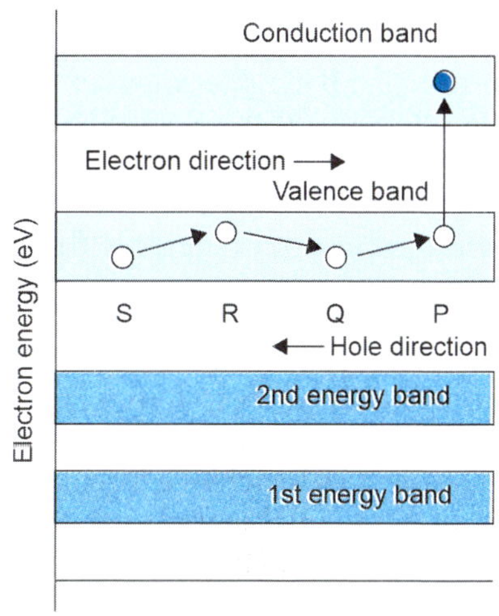

Fig. 3.6: Hole current in band diagram

3.6 TYPES OF SEMICONDUCTORS

As we know the semiconductor properties are lies in-between conductor and insulator. Depending upon the working and conductivity, semiconductors are classified into two types:

i. Intrinsic semiconductors, and
ii. Extrinsic semiconductors

3.6.1 Intrinsic Semiconductor

A semiconductor in very pure form is called as **intrinsic semiconductor**.

In a room temperature, hole and electron pairs are created in intrinsic semiconductor. When a small amount of stimulus or form of energy or electric field is applied, the current conduction takes place in two manners, namely free electrons and holes as shown in Fig. 3.7. It is discussed (*refer* to Section 3.2.), the free electrons are produced by thermal energy through a material that cause breaking of covalent bonds. Simultaneously, ejecting of electrons in material that form holes. So that, conduction of current through semiconductor is mainly due to both free electrons and holes.

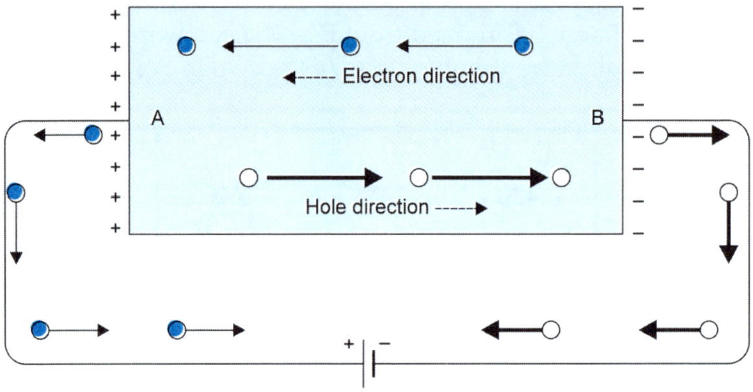

Fig. 3.7: Intrinsic semiconductor

It is also noted that current conduction in the circuit purely by electron in an external wire. Then what is the role of holes? From Fig. 3.7, it is cleared that hole being positively charged shift towards the negative terminal of battery. As it is reached to the negative terminal at B in material where it is joined with electrons that comes in opposite direction. Simultaneously, the electrons approach near the positive terminal at **A** are attracted away from their atoms into the positive terminal. This creates new holes near the positive terminal which again drift towards the negative terminal. In this way, hole and electrons conduct current in the pure form of semiconductor called intrinsic semiconductor.

3.6.2 Extrinsic Semiconductor

In pure type semiconductor (intrinsic) has a capability of conducting small amount of current only at room temperature. But it has scarcer practices in the field of electronics. To increase the conducting properties of the material, it should be added with some form of impurities to achieve its conductivity significantly. This can be done by adding a small amount of suitable impurity to a semiconductor. This type of material is called extrinsic semiconductor. The process of adding an impurity to a semiconductor is called *doping*. The ratio of impurity to atom is very low, i.e. 1 atom of impurity for every 100 to 1000 atom of semiconductors. When impurity added to atom will not affect the crystal structure of the semiconductor significantly.

Generally, impurities of semiconductor crystals are trivalent or tetravalent elements. Depending upon the impurity of semiconductors are classified in two types:

i. n-type semiconductor
ii. p-type semiconductor

If a pentavalent impurity (having 5 valence electrons) is added to semiconductor that create a large number of free electrons in the semiconductor. This type of crystal is called n-type semiconductor. Conversely, a trivalent impurity (having 3 valence electrons) is added to semiconductor that create a large number of holes in the semiconductor crystal. This is called p-type semiconductor.

3.6.2.1 n-type semiconductor

When a small amount of pentavalent impurity is added to a pure semiconductor, it is known as *n*-type semiconductor. The addition of pentavalent impurity provides a large number of free electrons in the semiconductor crystal. For example, arsenic (Z=33) and antimony (Z=51) whose valence electrons are 5. Such type of impurities is called donor impurities because it provides free electrons to the semiconductor crystal. These 4 of 5 electrons are in covalent bond with semiconductor atom and 1 remaining electron produces negative charge for the atom. So, it is called n-type semiconductor.

To explain the concept of formation of n-type semiconductor, consider a pure silicon crystal. As we know that a pure silicon atom has four valence electrons and it is added with a small amount of pentavalent impurity like arsenic. More number of free electrons are made available in the crystal during doping.

We can understand the concept clearly, for example, arsenic is pentavalent which means atom has five valence electrons. When this arsenic atom is bonded with silicon crystal in such a way that its four valence electrons form covalent bonds with four silicon atoms. The fifth valence electron of arsenic atom could not have chance to bond with any silicon atom and it is free which is shown in Fig. 3.8. Therefore, for every arsenic atom added, one free electron is made available in the silicon crystal. So that extremely small amount of arsenic impurity provides enough atoms to supply millions of free electrons to conduct current.

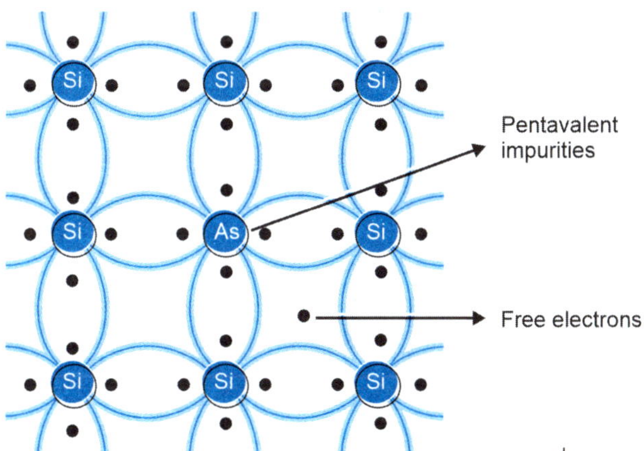

Fig. 3.8: Hole current in band diagram

The concept of free electron formation in n-type semiconductor can understand easily in the energy band description as shown in Fig. 3.9. Many free electrons are produced by the addition of pentavalent impurity that is responsible for the conduction of current by shifting it from valence band to conduction band. In addition, at room temperature thermal energy also generates a few hole-electron pairs. The current conduction in an n-type semiconductor is predominantly by free electrons, i.e. negative charges and is called n-type or electron type conductivity.

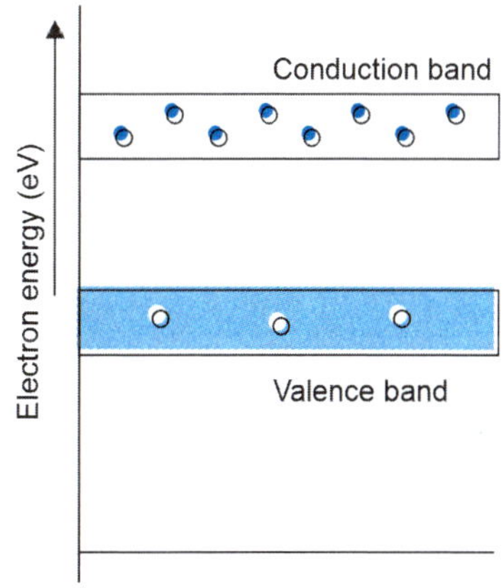

Fig. 3.9: Band energy of n-type

If potential difference is applied to *n*-type semiconductor, the free electrons are attracted towards the positive terminal of battery that constitute current (*refer* to Fig. 3.7). The current flow in the circuit is mainly due to electron which carries negative charge and it is called n-type conductivity.

3.6.2.2 p-type semiconductor

When a small amount of trivalent impurity is added to a pure semiconductor, it is known as p-type semiconductor. The addition of trivalent impurity provides a large number of holes in the semiconductor crystal. For example, gallium (Z=31) and indium (Z=49) whose valence electrons are 3. Such type of impurities is called acceptor impurities because it provides the holes that can accept the electrons. There are 3 electrons in the valence shell of trivalent atom are engaged covalent bond with semiconductor atom Si or Ge. So, it is formed 3 covalent bonds. In the fourth bond, only silicon atom contributes on valence electron but impurity does not have electron for bond and it is incomplete with shortage of one electron (Fig. 3.10). This shortage of electron is called a hole. In addition of trivalent impurity has produce a large number of holes (in millions) which have a deficit of electron carries positive charge. At room temperature thermal energy also generates a few hole-electron pairs. The current conduction in p-type semiconductor is predominantly by hole and is called p-type or hole-type conductivity.

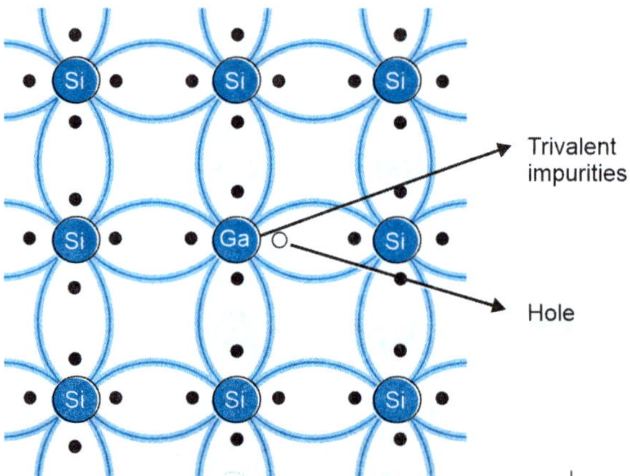

Fig. 3.10: p-type semiconductor crystal

The concept of hole created in p-type semiconductor can understand easily in the energy band description as shown in Fig. 3.11. If potential difference is applied to p-type semiconductor, the holes are attracted towards the negative terminal of battery by shifting electron towards positive terminal that constitute current (Ref. Fig. 3.7). The current flow in the circuit is mainly due to electron moves from one covalent bond atom to other that shifted hole in opposite direction. It is called p-type conductivity.

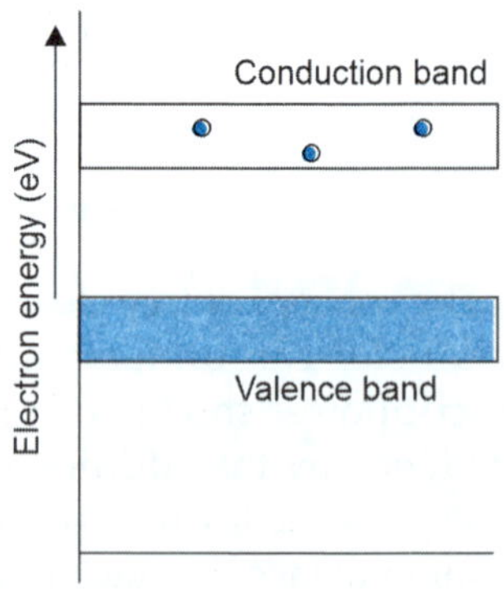

Fig. 3.11: Band energy of p-type

3.6.2.3 Carriers of n-type and p-type semiconductors

As we know, in a pure type semiconductor is doped with impurities that forms extrinsic semiconductor. On the effect of impurity, in extrinsic semiconductor n-type material has a large number of free electrons whereas p-type material has a large number of holes.

At room temperature covalent bonds of atom breaks and releases an equal number of free electrons and holes pairs. The current conduction through n-type material is mainly due to the flow of electrons and n-type material has large no free electrons which is considered as majority carriers. Conversely, the current conduction through p-type material is mainly due to the large number of holes created by doped trivalent atoms. Hole is majority carriers in p-type semiconductor.

NOTE

i. The movement of holes in semiconductor is deprived as it is more bound with nucleus.
ii. The availability of electrons in the valence band has a capability to jump into conduction band which is more movable than holes available in the valence band in a p-type semiconductor.
iii. The electron-hole pairs are formed at room temperature. But the free electrons availability is in minute quantity and also can carry a little amount of current in the p-type semiconductors.

3.7 P–N JUNCTION

Definition: A p–n junction is type of semiconductor and it is boundary between two semiconductors, namely a p-type and n-type semiconductors. When p-type is joined to the n-type, the contact surface is called *p–n junction*.

As a semiconductor in the pure form is called intrinsic semiconductor and impure form is called extrinsic semiconductor and extrinsic form is further differentiated into n-type and p-type. All these semiconductors have a minimum level of usage in the field of electronics. But it creates a revolution in the field of semiconductor when it is joined together, called junction.

In the field of semiconductors, it may also have one or more p–n junctions. It has a great importance in control element of semiconductor devices. If we have a basic knowledge on formation and properties of p–n junction and we can understand the other conducting properties of semiconductor devices easily.

Construction and properties of p–n junction

In general, a p-type semiconductor is joined to n-type to a make a p–n junction under certain conditions. After joining of semiconductor materials, electrons which is more in n-type and hole which is more in p-type that diffuse between boundaries until it gets saturated. It will create boundary which inhibits the scattering and movement of electron and holes.

In actual practice, the formation of p–n junction is done by adding a trivalent impurity to n-type silicon block and heated about 500 °C. This trivalent material and silicon melts and combined to form molten mixture. Later the temperature is lowered and started to become harden. Under suitable condition, the trivalent impurity will be adjusted to silicon. So that, the n-type silicon will be formed a junction with p-type appropriately.

Properties

After the formation of p–n junction, the electrons from n-type semiconductor reaches to the junction and diffuse into the p-type region where it will combine with holes and leaves one positive charge ion in n-type semiconductor. This results in loss of electron in n-type region and joined with holes in p-type region become neutralized. This creates a positive layer which is left by electron near the junction. Similarly, holes from p-type reaches to the junction where it combined with electron in the region and left its ionized electron in p side. This results in formation of layer of negative near the junction. These two layers are jointly formed a region is called *depletion region* or *layer* as shown in Fig. 3.12. This depletion region has potential that prevent the further movement or joining of electron and holes.

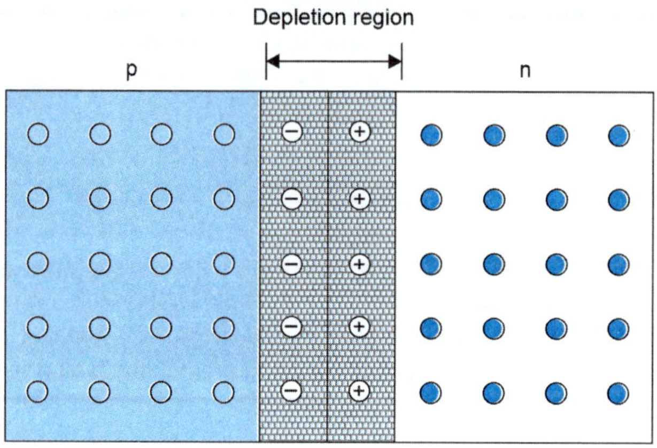

Fig. 3.12: p–n junction

3.8 EFFECT OF DC VOLTAGE ON P–N JUNCTION

The following are three biasing conditions of p–n junction diode which is based on the applied voltage,

i. **Zero bias:** If the p–n junction diode is not applied with voltage is called zero bias.
ii. **Forward bias:** If the positive terminal of the battery is connected to the p-type of semiconductor, whereas the negative terminal is connected to the n-type of semiconductor which constitutes **forward bias**.
iii. **Reverse bias:** Conversely, the positive terminal of the battery is connected to the n-type of semiconductor whereas the negative terminal of the battery is connected to the p-type of semiconductor which constitutes **reverse bias**.

3.8.1 Forward Biasing of p–n Junction

When a 'dc' voltage is applied to p–n junction, there is a passage of current in p–n junction as the potentials of battery cancels the potential barrier. Thus, permitting current flow, it is called forward biasing as shown in Fig. 3.13.

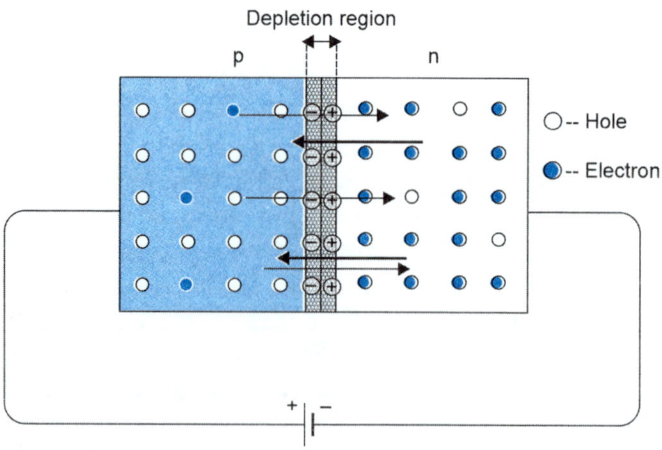

Fig. 3.13: Forward biasing

In a p–n junction, p-type is connected to the positive terminal of the battery and n-type is connected to negative terminal of the battery. Then p–n junction is constituted forward biased. When p–n junction at this state conducts current, the applied electric field is opposite to built-in electric field, i.e. depletion region potential. The electric field in the junction become weak and depletion region is reduced as shown in Fig. 3.13. The potential barrier voltage is in range between 0.1 to 0.3V which is very small. The external applied voltage is eliminating the barrier potential as forward biasing and junction has zero resistant. Therefore, current flows in the circuit. This is called forward biasing of p–n junction.

3.8.2 Reverse Biasing of p–n Junction

When a 'dc' voltage is applied to p–n junction, there is no flow of current in p–n junction as it is supporting the potential barrier. Thus, barrier potential increases which is not permitting current flow, it is called reverse biasing as shown in Fig. 3.14.

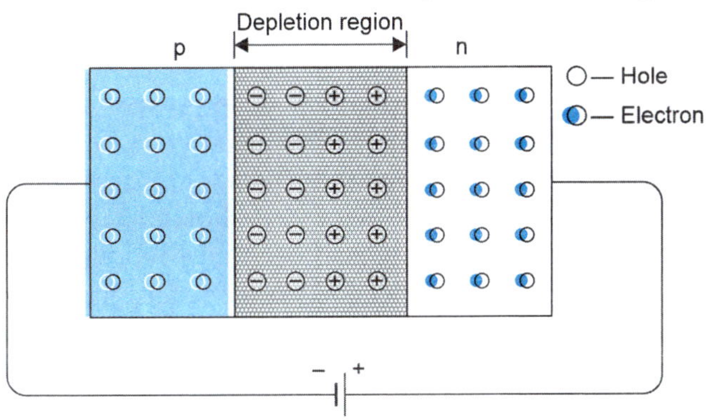

Fig. 3.14: Reverse biasing

In a p–n junction, p-type is connected to the negative terminal of the battery and n-type is connected to positive terminal of the battery. Then p–n junction is constituted reverse biased. When p–n junction at this state will not conduct current, the applied electric field is supported the barrier potential which causes to increase depletion region as shown in Fig. 3.14. When p–n junction in reverse bias, the free electrons from n region in attracted by positive terminal of battery and holes might be shifted to negative terminal which cause the deletion region wider and barrier potential is also increased which avoid the flow of electrons. Thus, no current during reverse bias.

3.8.3 Potential Difference—Electron Flow Characteristics of p–n Junction

Potential difference is nothing but voltage with volt (V) as unit and electron flow is the current with unit (amp) has characteristics in p–n junction (also called p–n junction diode). The characteristics of p–n junction diodes is a curve between V and I through the circuit. The known value will be taken in x-axis and dependent value will be plotted against y-axis. Voltage is known value and it is taken in x-axis while current is taken along y axis. The following graph is VI characteristics curve of p–n junction diode (*refer* to Section 3.8). We can understand the diode works under three biases, namely zero, forward and reverse biases.

i. When there is no external voltage is applied to the p–n junction and the junction diode is at under zero bias. So that potential barrier will not allow the flow of current as it is already in saturated depletion region.

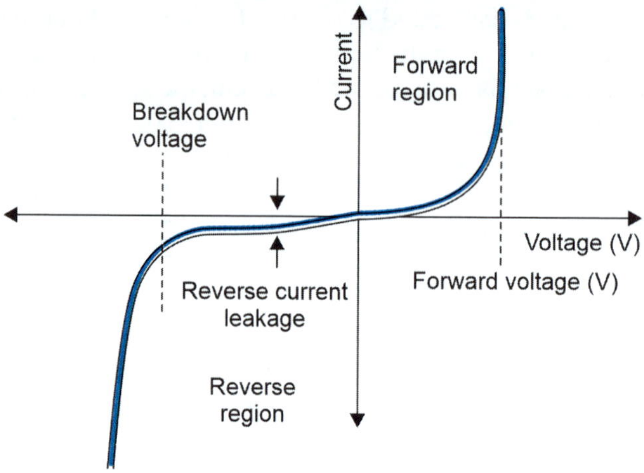

Fig. 3.15: VI characteristic curve

ii. When the p–n junction diode is forward bias, the p-type is connected to positive terminal of battery and n-type is connected to negative terminal. When the voltage of 0.3 to 0.7 V is supplied, depletion region decreases gradually and there is a gradual increase in current flow and become linear that showed in graph as forward region.

iii. When the p–n junction diode is reverse bias, the p-type is connected to negative terminal of battery and n-type is connected to positive terminal. This results in increase the depletion region which resists more to the flow of current in the circuit. But under certain condition a trifling current flow is observed due to some minority carrier leakage in the barrier. The reverse leakage current is shown in Fig. 3.15.

Incessantly, if reverse voltage is increased which can cause increase energy of electrons that forced to eject electron from semiconductor. This causes breakdown of junction and sudden rise of reverse current will be detected and damaged the junction eternally.

3.9 SEMICONDUCTOR DIODE AS CRYSTAL DIODE

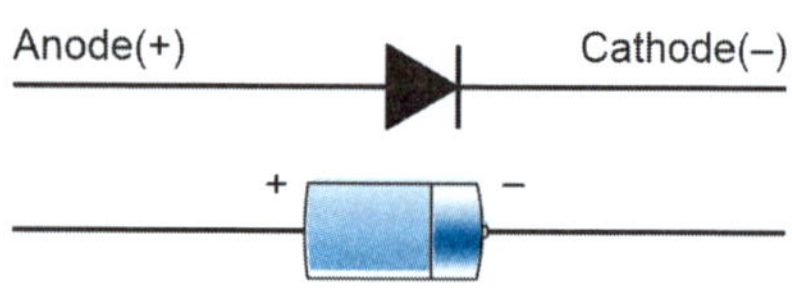

Fig. 3.16: Semiconductor diode and symbol

Semiconductor diode is principally called p–n **junction diode**. It can also be called crystal diode because it is made up of doped Si or Ge or other type semiconductor crystal only.

The principal role of p–n junction diode is conduct current only in one direction which means forward bias. So, it as an important role in the field of electronics as *rectifier*. The symbol of diode is shown in Fig. 3.16.

The p–n junction diode is formed in such a way that it has two terminal which allow the flow of current in one direction or no current. It is depending upon nature of connection like forward or reverse bias. If the diode is connected as forward bias, then it has high conductivity and very low resistance to the flow of current. Similarly, if the diode is connected as reverse bias, then it has low conductivity with very high resistance to current flow.

3.9.1 Rectifier Diode

Rectifier diode is nothing but a semiconductor diode. It is used to rectify an alternating current (AC) to direct current (DC) with help of diode. It can be used to connect any part in the circuit where the current pass-through diode. The diode which is made up of silicon have a forward voltage drop of 0.7 V and it has capable of conducting high electric current values. The voltage drops across the load is proportional to the current that available to convert. But in case of Germanium diode has a benefit as compared to silicon diode that it has low threshold voltage when diode is forward bias. This diode has capable to conduct the current which changes from mA to a few kA and voltages up to a few kV.

3.9.2 Classifications of Rectifiers

Rectifiers can be classified based on alternating current which is taken as input. These are:

a. Single-phase rectifiers
b. Polyphase rectifiers: It may be also classified in to two categories depending upon the period of conduction namely
c. Half-wave rectifiers
d. Full-wave rectifiers

Full wave rectifiers may be further classified into two categories depending upon nature of the circuit connection. It has two types namely

- Centre tapped full-wave rectifier
- Bridge full-wave rectifier

3.9.2.1 Single-phase rectifiers

Single-phase rectifiers, the name itself it is understood that the rectifier has an input of one single phase AC. The circuit is quite simple, it needs one, two or four diodes depend upon nature of circuit connection as shown in Fig. 3.17. The single-phase rectifier delivers a small amount of power. It uses only single phase in transformer for the conversion and rectifier diode connected in secondary coil for single phase transformer. The disadvantage of this rectifier is high ripple factor which means fluctuating of additional AC components that will be in the resolved output.

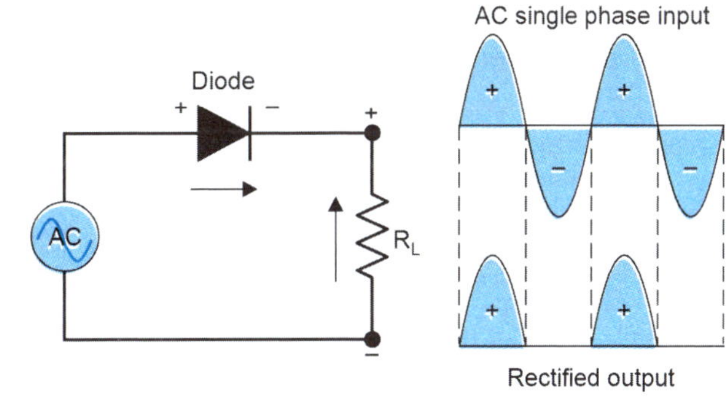

Fig. 3.17: Single-phase rectifiers and wave form

3.9.2.2 Polyphase rectifiers

Polyphase rectifiers which mean three phase rectifiers have an input of 3 phase AC. The circuit structure needs three or six diodes and these are connected to each phase of secondary coil of transformer. To reduce ripple factor three phase rectifiers can be used instead of single-phase rectifier. When comparing these rectifiers of alternating voltage category, three phase rectifiers is preferred while using in large systems. This due to delivering large amount of power and not required any addition filter to avoid fluctuation. It is more efficient and have more transformer utilization factor (TUF) which is defined as the ration of power delivered to the load and AC rating of the secondary transformer.

The three-phase rectifier working description is discussed as follows. Three phase supply is simply 3 single phases are joined together as shown in Fig. 3.18. All the three phases are having 120° in phase difference each other. This three-phase input signal is given to rectifier which produced rectified DC output voltage.

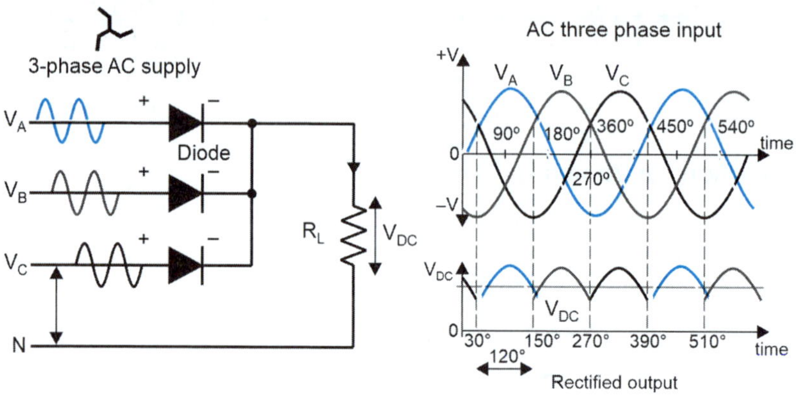

Fig. 3.18: Three-phase rectifiers

Comparing the two types of rectifiers, three-phase has advantage over single-phase because it can used in large systems that delivered large amounts of power and do not require any additional filter to reduce ripple factor. Because of this, three phase rectifiers are more efficient and have more transformer utilization factor.

3.9.2.3 Half-wave rectifiers

In half-wave rectification, the rectifier conducts the input AC current only during the positive half-cycles. It will cut-off the negative half-cycles

Construction: The half-wave rectifier circuit diagram is shown in Fig. 3.19, where, T is a transformer. The primary coil of transformer is supplied with AC input. The secondary coil of transformer is connected with diode D and load resistance RL in the circuit.

Working: The supplied AC input voltage produce a changing flex in NP which is linked to the secondary coil NS and this changing flex in the secondary coil is produce a current in the circuit. During first positive half cycle of AC input in the secondary coil. The point A is positive and B is negative, during first half cycle, the diode is forward biased. Hence, there is current flow in the circuit and there is DC output voltage through a load resistance RL. Next during negative half cycle, the point A is negative and the point B is positive. So, the diode D is reverse biased. Hence, there is no current flow in the circuit and no DC output voltage. Therefore, current always flows in one direction only that we can see in the wave form.

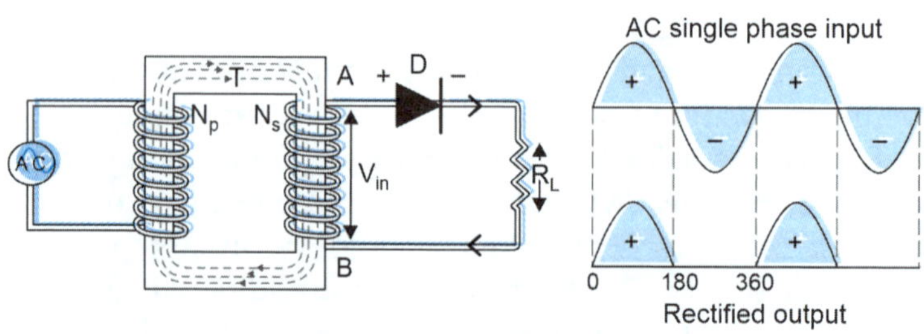

Fig. 3.19: Half-wave rectifiers

Peak inverse voltage (PIV): It is defined as the type of reverse voltage in which maximum voltage is withstand by rectifier diode during the non-conducting period. Non-conducting period means a diode can withstand in the reverse direction without breaking down.

Disadvantages: The disadvantages of a half-wave rectifier are:

i. The output is low which means input AC supply delivers powers only half as DC output.

ii. The pulsating current as DC output also contains alternating component whose frequency is equal to the supply frequency. Therefore, suitable filtering is required to produce steady direct current.

3.9.2.4 Full-wave rectifiers

Full-wave rectifier is defined as a type of semiconductor diode rectifier that converts both halve waves of alternating current signal into a direct current signal. The construction of full wave rectifier circuit, full wave rectifier is required with multiple diodes. There are two types of full-wave rectifiers namely,

i. Centre tapped (two diode) rectifier, and

ii. Bridge rectifier

3.9.2.4.1 Two diode or centre tapped rectifier

The centre tapped rectifier circuit consists of two diodes D_1 and D_2 which are connected to the secondary coil of transformer. According to Farday's law of electromagnetic induction, the AC input signal is inducted to secondary coil which given as input for rectifier.

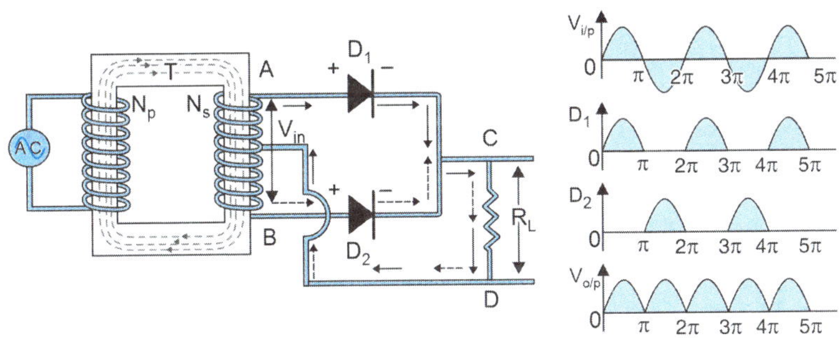

Fig. 3.20: Centre tapped full-wave rectifiers

Construction: The full-wave rectifier circuit diagram is shown in Fig. 3.20, where, T is a transformer. The primary coil of transformer is supplied with AC input. The secondary coil of transformer is connected with two diode D_1, D_2 and load resistance R_L in the circuit.

Working: The supplied AC input voltage produce a changing flux in primary coil, NP which is linked to the secondary coil N_S and this changing flux in the secondary coil is produce a current in the circuit. During first positive half cycle of AC input in the secondary coil one end of the secondary at A becomes positive and at B becomes negative. So that the diode D_1 is forward biased whereas the diode D_2 is reverse biased. As a result of this the D_1 conducts current and D_2 does not conduct current. There is an output through load resistance R_L, the input and output wave forms are shown in Fig. 3.20.

During negative half cycle of AC input in the secondary coil one end of the secondary at A becomes negative and at B becomes positive. So that the diode D_1 is reverse biased whereas the diode D_2 is forward biased. As a result of this, the diode D_2 conducts current and the diode D_1 does not conduct current. There is an output through load resistance RL and the output wave form D_2 is shown in Fig.3.20.

Therefore, during both of half cycles, there is a current flow through the load resistance. The output is developed in both cycles of AC input which pulsating DC output. The full wave forms are rectified.

Peak inverse voltage (PIV): PIV is the maximum reverse voltage where the rectifier has to withstand. In non-conduction period, if diode D_1 is conducting and D_2 non-conducting and vice versa.

3.9.2.4.2 Bridge rectifier circuit

The bridge rectifier in which four or more diodes are combined to form a bridge circuit and it is efficiently converting alternative current to direct current. The advantage of the rectifier is more efficient in rectifying AC current when compared with all other rectifiers.

Construction: The full wave rectifier circuit diagram is shown in Fig. 3.21, where, T is a transformer. The primary coil of transformer is supplied with AC input. The secondary coil of transformer is connected with two diode D_1, D_2, D_3 and D_4 which are arranged in balancing two legs of a bridge circuit in the form of wheatstone bridge system and also with load resistance R_L in the circuit.

Working: When an alternative current is applied to the primary coil of transformer N_p, it will produce current in secondary coil N_s according to the law of induction. During positive half cycle in secondary coil at C is positive and at D is negative. Diode D_3 and D_4 are forward bias and the current flow through the direction ACFR$_L$EDB and at the same time diode D_1 and D_2 are reverse bias.

During negative half cycle in the circuit at C is negative and D is positive. Diode D_1 and D_2 are forward bias and the current flow throught the direction ACER$_L$FDB and diode D_3 and D_4 are reverse bias. Therefore, during both halves of cycle, there is an output voltage through R_L. So that full wave of input is rectified and output waveform is shown in Fig. 3.21.

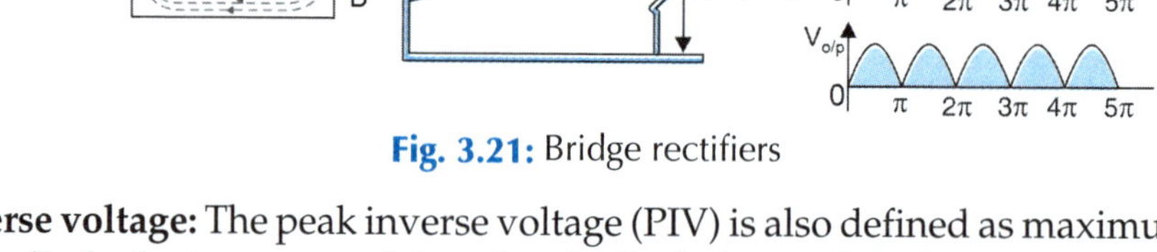

Fig. 3.21: Bridge rectifiers

Peak inverse voltage: The peak inverse voltage (PIV) is also defined as maximum possible voltage in diode during reverse bias of each diode is equal to the maximum secondary voltage of transformer and it is one half of PIV of center tapped.

Advantages

i. It can eliminate centre tapped transformer for better efficiency.
ii. The output is doubled than centre tapped circuit.
iii. The PIV is one-half that of the centre tap circuit.

Disadvantages

i. It requires four diodes.
ii. Voltage drop is more as it used four diodes.

3.10 TRANSISTORS

The name transistor referred as combination of two words transfer and resistor. Basically, a transistor is made up of three pieces of semiconductor device. Specifically extrinsic semiconductor in that a combination of three alternate with two junctions of p-type and n-type semiconductors. It is like a sandwich of 2p-type with n-type or 2 n-type with p-type semiconductor.

It is an electronic component and can be found in different circuits. It is also used in digital electronics in amplifying, high power application in that on/off switch actions, etc.

In electronics, there are different types of transistor which are specialized in its own application. These are:

i. Bipolar junction transistor (BJT)
   a. NPN and b. PNP
ii. Field effect transistor (FET)
   a. JFET and b. MOSFET
iii. Unijunction transistor

3.10.1 Bipolar Junction Transistor (BJT)

BJT has two PN junctions and it is also called normal transistor. It has two configuration NPN and PNP. Usually, NPN transistor is preferred because it has electron as a majority carrier whereas PNP has holes as majority carrier. As we know, the electron has better mobility than holes.

NPN transistor is formed by placing a p-type semiconductor in-between two n-type semiconductors. Similarly, the PNP transistor also formed in a same manner like placing n-type semiconductor between two p-types.

Configuration: It is a three terminal device which are having three possible ways to connect in an electronic circuit. These three terminals are being named as base, emitter and collector. One terminal in common to both input and output. This can be formed by connecting two diodes back-to-back. So that it has two p–n junctions. The construction of transistors is shown in Fig. 3.22.

Three terminals are taken from the transistor which indicate emitter, base and collector respectively.

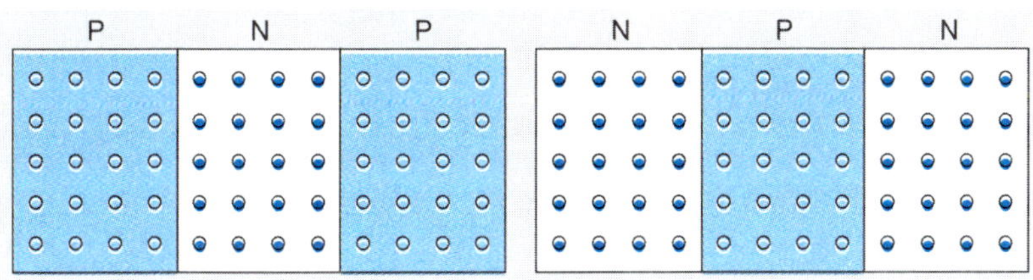

Fig. 3.22: Construction of PNP and NPN transistor

Emitter

- Usually, emitter should be doped heavily. It has moderate size with main function as to supply majority carriers.
- To identify better, it will be in left hand side (Fig. 3.22).
- Principally, it emits electrons and named as emitter.
- It is indicated with 'E' letter.

Base

- Usually, base should be doped lightly. It has thin size with main function as to pass majority carriers from emitter to collector.
- To identify better, it will be in middle portion (Fig. 3.22).
- The main function of base is to facilitate majority carriers from E to C.
- It is indicated with 'B' letter.

Collector

- Usually, collector should be doped moderately. It has size that should be larger than emitter and collector with function as to collect carriers.
- To identify better, it will be in right side (Fig. 3.22).
- Principally, it collects carriers and named as collector.
- It is indicated with 'C' letter.

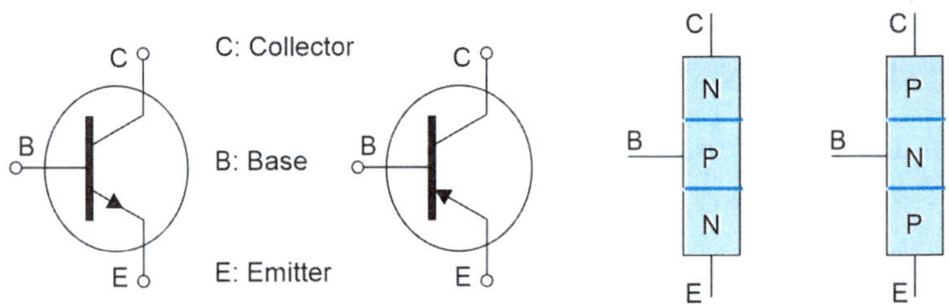

Fig. 3.23: Circuit and electronic symbol of transistor

Figure 3.23 shows the circuit and electronic symbol of NPN and PNP transistor. The arrow head in the transistor shows the emitter of transistor. The large size in transistor is collector which dissipate more power. Both emitter and collector have specific function in electronic circuit so it is not interchangeable.

Connections: The three-terminal device of transistor has three possible way to connect in electronic circuit. The one of three terminal in transistor is being common to both input and output. Thees are:

- **Common base:** It has voltage gain and no current gain.
- **Common emitter:** It has both voltage and current gain.
- **Common collector:** It has current gain and no voltage gain.

3.10.2 Field Effect Transistor (FET)

An FET is a semiconductor unipolar device with three-terminals. Unlike BJT it has a voltage-controlled device which are having high efficiency, instant operations and can be used in most electronic device. The main advantage of FET is with their low power consumption and power dissipation, very high input impedance and low heat dissipation.

The FET is a unipolar device, which means that it is made using either p-type or n-type material as main substrate. The current conduction of a BJT is mainly due to both majority and minority carriers contribute to current whereas in FET is done by either electrons or holes. There are two types of field effect transistors namely.

i. Junction field effect transistor (JFET)
ii. Metal oxide semiconductor field effect transistor (MOSFET)

Junction field effect transistor (JFET): JFET is a three-terminal device in which current conduction is done by any one type of carriers, i.e. electrons or holes. A JFET consists of p-type or n-type silicon bar with two p–n junctions as shown in Fig. 3.24. If it is made up of n-type then it is called n-channel JFET as shown in Fig. 3.24a and if it is made up of p-type then in called p-channel JFET as shown in Fig. 3.24b. The two p–n junctions forming diodes are connected with common terminal called gate (G). Other two terminal are called source (S) and drain (D).

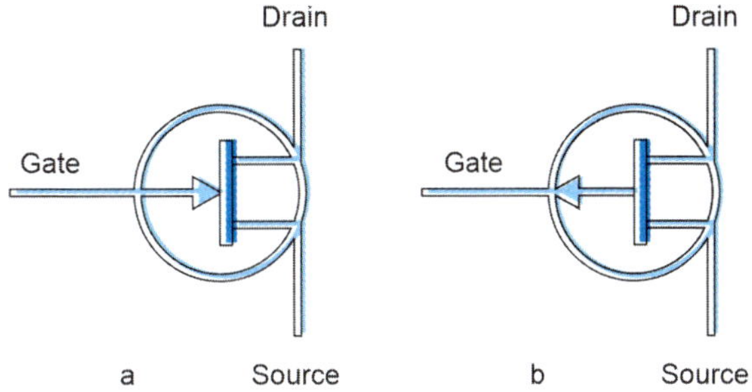

Fig. 3.24: (a) n JFET, (b) p JFET terminals

Metal-oxide-semiconductor field-effect transistor (MOSFET): MOSFET is also called insulated-gate field-effect transistor (IGFET) and it is due to gate input of transistor is electrically insulated from current carrying main channel. It is a voltage-controlled field effect transistor in that metal oxide electrode is electrically insulated. A very thin layer of insulating material like silicon diode is used as insulator of gate electrode.

As the gate terminal is electrically isolated from the main current carrying channel between the drain and source. So, no current flows into the gate which mean it acts like JFET of a voltage-controlled resistor. The input voltage of main channel is proportion to the current flow.

If both positive current and negative current are applied to MOSFET as main current on the gate which is already insulated from the channel. So that with negative gate bias voltage, it acts as a depletion MOSFET, whereas with positive gate bias voltage, it acts as enhancement of MOSFET.

Classification of MOSFETs: It is classified into two types based on the type of operations, namely enhancement mode MOSFET (E-MOSFET) and depletion mode MOSFET (D-MOSFET), these MOSFETs are further classified based on the material used for construction as n-channel and p-channel. So, in general, there are 4 different types of MOSFETs.

i. N-channel depletion mode MOSFET
ii. P-channel depletion mode MOSFET
iii. N-channel enhancement mode MOSFET
iv. P-channel enhancement mode MOSFET

The N-channel MOSFETs are called NMOS and they are represented in Fig. 3.25.

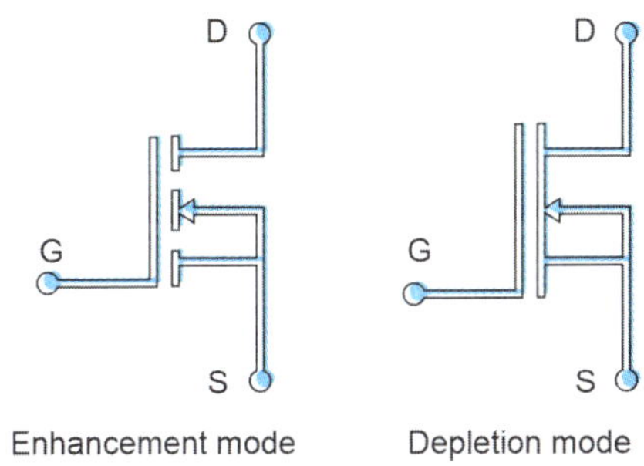

Fig. 3.25: NMOS

The P-channel MOSFETs are called PMOS and they are represented in Fig. 3.26. According to connections of MOSFET, the gate (G), drain (D) and source (S) are physically connected in depletion mode of MOSFET whereas as they are separated physically in enhancement mode. So that the symbol are shown broken for enhancement mode.

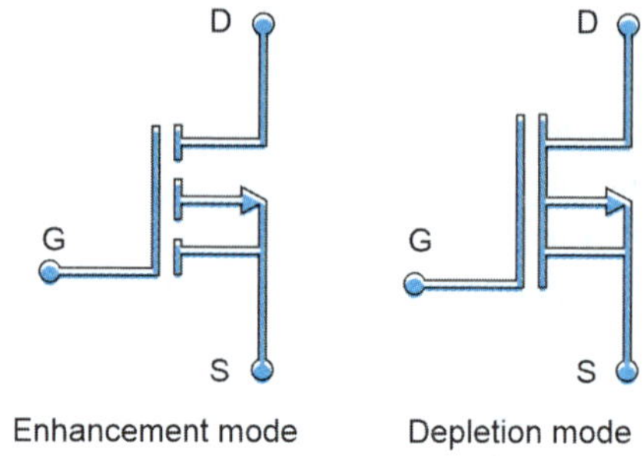

Fig. 3.26: NMOS

3.11 FILTER CIRCUITS

In electronic circuit, an essentiality of rectifier is a filter that produce a pure DC supply. The rectified output carries pulsating character which mean it contains AC and DC components. The AC is avoidable and should be removed from the load. To do this, a filter circuit should have a capability of removing the AC component and at the same time DC component should reach the load. To achieve this, the filter should have an inductors (L) and capacitor (C). Sometimes the filter in circuit is called LC coupled filter.

From the above, the filter circuit is defined as "the device which remove AC component in rectifier output but allows the DC component to reach the load".

Working: A filter circuit should be introduced between the rectifier and the load as shown in Fig. 3.27. It is combination of L and C and the actions are depending on electrical principles. Capacitor has a capability to pass AC but it does not allow DC. Conversely, an inductor opposes AC but allows DC to pass through the filter. So the filter should be arranged with suitable network of L and C which can effectively remove the AC component and allowing the DC component to reach the load.

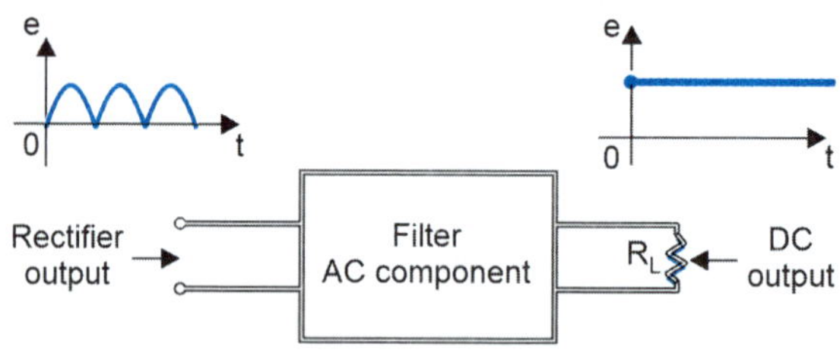

Fig. 3.27: Filter circuit

Types of filter circuits: There are three most commonly used filter circuits,

i. Capacitor filter
ii. Choke input filter
iii. Capacitor input filter or, filter

3.12 OSCILLATORS

An oscillator is an electronic device which can generate alternating current of a desired frequency. It is also converting DC power to AC power at the load from the DC source. Oscillation is achieved through positive feedback mechanism in amplifiers.

The essential components of an LC oscillator are shown in Fig. 3.28.

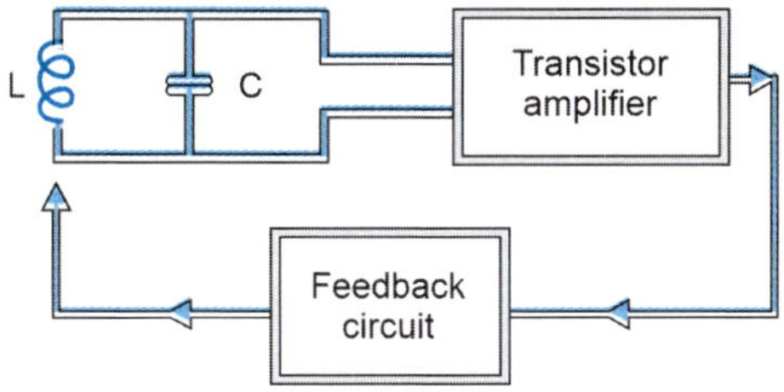

Fig. 3.28: Oscillator circuit

i. Tank circuit. Inductor L and capacitor C are connected parallel to the circuit. The values of inductance and capacitance can be determined by the frequency of oscillations in the circuit.

i.e., $f = \frac{1}{2\pi}\sqrt{\left(\frac{1}{LC}\right)}$

ii. Transistor amplifier. To amplify the oscillations produced by the tank circuit.

iii. Feedback circuit. A positive feedback network to transfer a part of output energy to the tank circuit in proper phase as input. The amount of energy feedback is enough to compensate for the losses of energy in the LC circuit.

Types of Oscillators

i. Armstrong oscillator
ii. Colpitts oscillators
iii. Cross-coupled oscillator
iv. Crystal oscillator
v. Dynatron oscillator
vi. Hartley oscillator
vii. Meissner oscillator
viii. Optoelectronic oscillator
ix. Phase shift oscillator
x. RC phase shift oscillator
xi. Robinson oscillator
xii. Tri-tet oscillator
xiii. Wine bridge oscillator

Chapter

4

Electricity

INTRODUCTION

In physics, electricity is a form of energy in which flow of electrons involves in changing its kinetic energy. The energy is mainly depending on the speed of the charges, if it moves faster and it carries more amount of electrical energy. As we know, all the matters of the universe are made up of an atom and it has nucleus at the center. The nucleus has positively charged matters called protons and uncharged matters are neutrons. The nucleus is enclosed by negatively charged matters called electrons. Whenever an external force in act on an atom that disrupts binding energy between protons and electrons, so an atom may gain or lose an electron and the atom gets ionized. When the electron lost from an atom, it is free to move in the conductor that causes an electric current.

4.1 ELECTRIC CHARGE

Electric charge, some particles are affected either by an electric or magnetic field which carries charge that are basic properties of matter. It may be positive or negative which are two general types of charge with natural units and is neither created nor destroyed.

All experimental evidence showed that electric charge in nature comes in units of magnitude only. It may be denoted as '*e*' and is also called electronic charge and naturally occurring charges are integral multiples of the basic unit, i.e. $q = ne$ where n is either positive or negative integer. So, the charge exists in discrete packets rather than continuous amounts. Charge is therefore quantized.

According the law of conservation of electric charge states that the algebraic sum of the electric charge remains constant in a closed system. This indicates that charge can neither be created nor destroyed.

4.2 CHARGE DENSITY

Charge density is nothing but accumulation of electric charge in a given unit. The unit dimension of amount of electric charge can be measured as follows:

i. **Charge per unit length:** If a charge q, distributed over a certain length that determined linear charge density. The SI unit is coulomb per meter (C/m).

$$\lambda = \frac{q}{l}$$

ii. Charge per unit surface area: If a charge q spread over a certain surface area that determined surface charge density. The SI unit is coulomb per meter square (C/m^2)

$$\sigma = \frac{q}{A}$$

iii. Charge per unit volume: If a charge q distributed over a certain volume that determined volume charge density. The SI unit is coulomb per meter cube (C/m^3)

$$\rho = \frac{q}{V}$$

The distribution of electric charge is determined the charge density and charge may be positive or negative. It may be measure of electric charge per unit area of a surface or per unit volume of a body or field.

It is also described how much of electric charge is accumulated in a particular field. Mainly, it finds the charge density per unit volume, surface area, and length.

It measures the amount of electric charge per unit measurement of the space. This space may be one, two or three dimensional. Charge density will depend on the position, which can be negative.

4.3 COULOMB'S LAW

The Coulomb's law states that *"the force of attraction or repulsion between two-point charges is directly proportional to the product of the charges and inversely proportional to the square of the distance between them"*.

Let us consider two-point charges q_1 and q_2 which are placed at A and B separated by a small distance r as shown in Fig. 4.1.

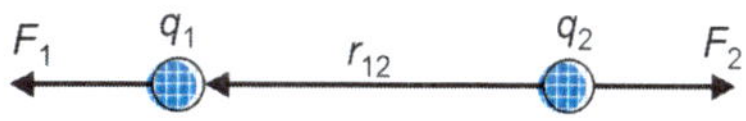

Fig. 4.1: Two-point charges

The vector F_1 is the force experienced by q_1, and the vector F_2 is the force experienced by q_2. If $q_1q_2 > 0$ the repulsive forces are experienced between charges and if $q_1q_2 < 0$ the attractive forces are experienced between charges. The magnitude of the forces will always be equal.

According to Coulomb, the magnitude of the electrostatic force F between two-point charges q_1 and q_2 is directly proportional to the product of the magnitudes of charges and inversely proportional to the square of the distance r between them. Like charges repel each other, and opposite charges mutually attract.

$$F \propto \frac{q_1q_2}{r^2} \quad ...(4.1)$$

$$F = K\frac{q_1q_2}{r^2} \quad ...(4.2)$$

where, $$K = \frac{1}{4\pi\varepsilon_0} \approx 9 \times 10^9\ N{\cdot}m^2/C^2 = \text{Coulomb's constant}$$

Therefore, $$F = \frac{1}{4\pi\varepsilon_0}\frac{q_1q_2}{r^2} \quad ...(4.3)$$

In Eq. (4.3), if $q_1 = q_2$ and $r = 1$, we have

$$F = \frac{1}{4\pi\varepsilon_0}\frac{q_1q_2}{r^2} = (9\times10^9)\frac{1\times1}{1^2}$$

$$F = (9 \times 10^9)\ N$$

The SI unit of charge is the coulomb.

A coulomb is defined as the quantity of charge in which equal and similar unit charges are placed at a distance of 1 meter in air or vacuum, experiences a repulsive force of (9×10^9) N.

A coulomb is also defined as the amount of charge that passes in 1 sec through a cross-section of a wire conducts 1 ampere current in the wire.

4.4 ELECTRIC FIELD

Let us consider a test object carrying a positive electric charge q that placed near a charged plate or rod, an electrostatic force F will act on it. So, there is an electric field in this region. It is denoted by a vector E

$$\therefore \qquad E = \frac{F}{q} \qquad ...(4.4)$$

From the above equation, it is defined as the electric field in the region at a point is the force that acts on a unit positive charge placed at that point. The SI unit for the electric field is the newton/coulomb (NC^{-1})

Similarly, the force F exerted on a charge q at a point is the product of the electric field E and the charge q from equ. (4.4) can be written as

$$\therefore \qquad F = \frac{q}{E} \qquad ...(4.5)$$

4.5 ELECTRIC POTENTIAL

The electric potential is also called voltage, which is the difference between two places in potential energy per unit charge in an electric field.

The electric potential at a point is defined as the amount of work done in moving unit positive charge from infinity to that point without change in velocity, against electrical forces.

If the potential difference between two points is said to be 1 volt, then 1 joule of work is done on moving charge of 1 coulomb from one place to other place where the charge movement is against electrical force.

Consider an isolated point of charge $+q$ is at O (Fig. 4.2). A and B are two points in electric field. Work done in moving a unit positive charge from A to B is W_{AB}

Fig. 4.2: Potential difference between two points

So $\qquad W_{AB} = V_B - V_A$

Where, V_A and V_B stand for the potential at A and B

If A is positioned at infinity, the $V_A = 0$.

Then $\qquad W_{AB} = V_B$

Here, W is the work done in moving a unit positive charge from infinity to the point B. V_B is the potential at B.

4.6 ELECTROMOTIVE FORCE

The electric potential produced by electrochemical or changing the magnetic field and it is also called electromotive force (EMF).

Electrochemical reaction on a generator or a battery is used for the conversion of energy from one form to another. Here, one terminal is positively charged whereas the other terminal negatively charged. Therefore, an EMF is an amount of work done on a unit electric charge. It is also called a type of potential.

In the electromagnetic flowmeter is also influenced by electromotive force and flowmeter is an application of Faraday's law.

Symbol of EMF is ε and the unit of EMF is volt.

Dimension of electromotive force: EMF is given as the ratio of work done on a unit charge which means equivalent to metre–kilogram–second system to one joule per coulomb of electric charge and it is represented as:

$M^1L^2T^{-3}I^{-1}$ (work done = $M^1L^2T^{-2}$ and electric charge = IT)

4.7 BATTERIES

A battery is said to be a power source which consists of one or more electrochemical cells with external connections. Two different metals are in contact with the help of free moving ion in a liquid constitute an electrochemical cell. A set of connected cells is called battery. There are two basic forms of battery that are primary and secondary. In primary cell, a chemical reaction that shoots the power of batter in which if chemicals are exhausted then the battery is completely ceased. In contrast, reversible chemical reactions are observed in secondary cell. In secondary cell, when the chemical reaction takes place impetuously, the battery produces a potential difference. From an external voltage supply is applied, the same potential difference is produced by the chemical reaction runs in reverse. A battery in which it made up of secondary cells is said to be a rechargeable battery.

Types of cells: There are many different types of electrochemical cells which are available with varying chemical processes and designs. The names of some are galvanic cells, electrolytic cells, fuel cells, flow cells and voltaic piles.

4.7.1 Zinc-carbon Batteries

Zinc-carbon batteries are being popular in providing an economical power source for devices for past 10 decades. In general, there are two types of zinc-carbon battery available that are:

i. Leclanche battery, and
ii. Zinc chloride battery

In 1866, Goerge Lionel Leclanche was invented this primary type. Ammonium chloride electrolytc was used in this battery in which it has low corrosive rate. Strong mineral acids were used as electrolyte in previous battery system.

Construction: In Leclanche battery, main container is a glass jar. It is filled electrolyte solution like ammonium chloride. A zinc rod immersed in electrolyte solution and it is used as negative electrode or anode which is already amalgamated. In this battery, a mixture of manganese dioxide and carbon powder is filled with porous pot. A carbon rod is placed in this mixture.

In the jar a carbon rod is placed in the ammonium chloride solution and it is served as positive electrode or cathode. Manganese dioxide and carbon powder are mixed with a resin gum blinder to provide a solid block. It created an advantage that a solid structure of cathode mixture is formed, so porous pot in not required for Leclanche battery.

This type of battery is refined by Dr Carl Gassner in 1988. Instead of liquid ammonium chloride he used a paste of plaster of Paris and ammonium chloride as electrolyte. In his battery the container itself made up of zinc, instead of glass container. Hence this container also serves as anode of the battery. Local chemical action can be minimized by wrapping zinc chloride – ammonium chloride saturated cloths to the cylindrical cathode mix block.The first commercial battery is designed with dry zinc carbon battery, in that gassner replaced plaster of Paris by wheat flour in electrolyte mixture.

After 1960, zinc chloride battery is popularly available commercial battery which is next version of zinc carbon battery. Here, zinc chloride is used as electrolyte instead of ammonium chloride which provides a better performance in the heavy drain application. In other words, zinc chloride battery is improved substitute of Leclanche' battery in heavy drain applications as shown in Fig. 4.3.

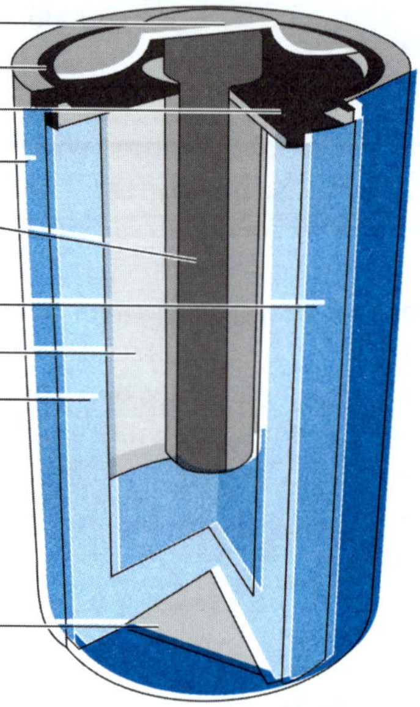

Fig. 4.3: Schematic diagram of zinc-carbon batteries

Chemical Reaction: $Zn + 2MnO_2 + ZnCl_2 + 2H_2O \rightarrow 2MnO(OH) + 2Zn(OH)Cl$

Advantages

i. Economical cost of the battery is quite low.
ii. Various shapes, sizes and capacities of these cells are easily available.
iii. Long reliability.

Disadvantages

i. Energy density is quite low.
ii. Poor service at low temperature.
iii. Poor leakage resistance.
iv. Poor perform at high current drain application.
v. Oxygen sensitive.
vi. High gassing rate

NOTE

In a device which absorbs energy of charge (such as recharging a battery), the cathode is negative (electrons flow into the cathode, and charge flows out of it), and in a device which provides energy (such as battery in use), the cathode is positive (electrons flow into it and charge flows out of cathode.

4.7.2 Lead-Acid Battery

This is a secondary type battery in which electrical energy can be stored in the form of chemical energy. Whenever required the chemical energy is converted into electrical energy.

The term here charging of battery is the conversion of electrical energy into chemical energy by applying external source of electrical energy. The discharging takes place by converting chemical energy into electrical energy to the load. The battery which uses sponge lead and lead peroxide for the conversion of the chemical energy into electrical power, such type of battery is called a lead acid battery. The lead acid battery is most commonly used in the power stations and substations because it has higher cell voltage and lower cost.

Construction: The parts of the lead acid battery are given below. The container and the plates are the main part of the lead acid battery. The container acts like a storage device which stores chemical energy and is converted into electrical energy with the help of the plates whenever it requires.

i. **Container:** The container of the battery is made of any one of lead lined wood, ebonite, glass, rubber of bituminous, ceramic materials or nowadays a molded good quality plastic. To avoid the discharge of electrolyte, it is placed at the top and the bottom of the container, there are four ribs, on two of them rest the positive plate and the others support the negative plates. To avoid the damage of electrolyte the material of container should be resistant to sulfuric acid.

ii. **Plate:** The plate of the battery is arranged in such a way that the cell is of diverse design and form like a grid which is made up of lead and the active material. The grid assembly is plays a important role in conducting the electric current and distributing the current equally on the active material.
Lead durability is increased with alloy of antimony which is placed at a right angle or diagonally because the positive and negative plates are of the same design and for the uniform conduction of the current.

iii. **Active material:** The active material is nothing but the active participation of plate in chemical reaction which means charging or discharging of electrical energy. The name of few active elements is:
   a. **Lead peroxide (PbO_2):** It forms the positive active material which is in blue in color.
   b. **Sponge lead:** It forms the negative active material which is orange in color.
   c. **Dilute sulfuric acid (H_2SO_4):** It acts as an electrolyte in which 31% of sulfuric acid present.

   The mechanical strength of the negative and positive active materials is so high and it can be used alone (Fig. 4.4).

iv. **Separators:** The non-conducting material can be used as separator and it is made up of chemically treated lead wood or rubbers or glass which is placed between the positive and negative to insulate them from each other.

v. **Battery terminals:** The positive and the negative plates/rods are used as battery terminals. The positive terminal with a diameter of 17.5 mm at the top is slightly larger than the negative terminal which is 16 mm in diameter.

Working principle of lead acid battery: The positive hydrogen ions ($2H^+$) and sulphate negative ions (SO_4^-) are formed when the sulfuric acid dissolves and move freely. If it is connected with the outsource of DC supply then the hydrogen ions are attracted by the negative terminal of the supply called cathode and SO_4^- ions are attracted by the positive terminal called anode as shown in Fig. 4.4.

If two-positive hydrogen ion takes two electrons from the cathode and negative sulphate ion takes the two ions from the anodes then react with water and form sulfuric acid and hydrogen gas. If it takes place the oxygen that produced from chemical reaction which ready to react with lead oxide and forms lead peroxide (PbO_2). Thus, during charging the lead cathode remain as lead, but lead anode gets converted into lead peroxide which is blue in color.

After the chemical reaction, the outsource DC supply is disconnected and if the voltmeter connects between the electrodes, it will show the potential difference between the electron that formed from chemical reaction. It can be confirmed by connecting it with wire, then current flows from the positive plate to the negative plate through external circuit, i.e. the cell is capable of supplying electrical energy.

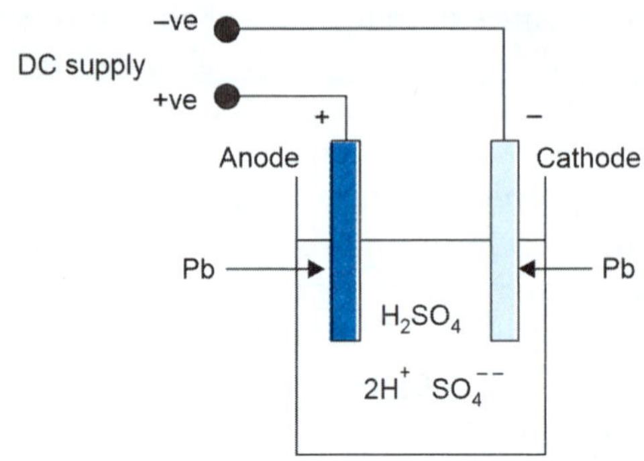

Fig. 4.4: Schematic diagram of lead acid batteries

Chemical action during discharging: During discharge of cell, electrons flow in a direction opposite to that during charging. When the loads are connected across the plates, the sulfuric acid again breaks into positive ions $2H^+$ and negative ions SO_4^-. The hydrogen ions react with the PbO_2 and make PbO and water H_2O. PbO starts reacting with the H_2SO_4 and creates $PbSO_4$ and H_2O.

On the other side SO_4^- ions exchange electrons from Pb, creating radical SO_4 which further creates $PbSO_4$ reacting with the Pb. The chemical reaction during discharging is as follows.

$$PbO_2 + Pb + 2H_2SO_4 \rightarrow 2PbSO_4 + 2H_2O$$

Chemical action during recharging: When the battery is connected to the charger of the DC supply, the sulfuric acid molecules break into ions, $2H^+$ and SO_4^-. The hydrogen ions being positively charged moved towards the cathodes and receive two electrons and form a hydrogen atom. The hydrogen atom reacts with lead sulphate cathode forming lead and form sulfuric acid according to the chemical equation. SO_4^- reacts with anode and become SO_4^-. This sulphate reacts with lead sulphate of anode and create the lead peroxide and sulfuric acid. This creates an energy gets stored by increasing the specific gravity of sulfuric acid and potential of the cell. The recharge reaction is as follows.

$$2PbSO_4 + 2H_2O \rightarrow PbO_2 + Pb + 2H_2SO_4$$

4.7.3 Alkaline Batteries

Alkaline batteries are commercially used in many things like in equipment, flash lights, etc. Usually it is used where there is irregular power supply, low voltage is required like watches, mobiles and many other appliances and electronic equipment. The main advantage of the battery is that it can be charged and used if the capability of supplying power reduces which is also called cell. Cells are the unit of battery; comprehensively a group of cells called battery.

The first alkaline battery was brought into market by Eveready battery, Toronto. It was developed by Lew Urry, a chemical engineer.

Construction of alkaline battery: The battery consists of a hollow steel drum in which all materials of the battery is placed and at the same time it act as cathode. The top of the drum is served as positive terminal of the battery which is projecting upwards. It consists of coal dust mixed with fine-grained manganese dioxide (MnO_2) powder to the inner peripheral surface of the empty cylindrical drum. This molded mixture serves as a cathode mixture of the alkaline battery. This mixture is covered with a paper separator which is placed inner surface of thick layer of cathode. The central space is filled with zinc powder with potassium hydroxide electrolyte which increases the contact surface and served as anode. A metallic pin (preferably made of brass) is inserted along the central axis of the alkaline battery to collect the negative charge. This pin is called negative collector pin. This pin is in touch with a metallic end sealed cap which is also covered by a plastic that separates positive steel drum and negative end cap of alkaline battery.

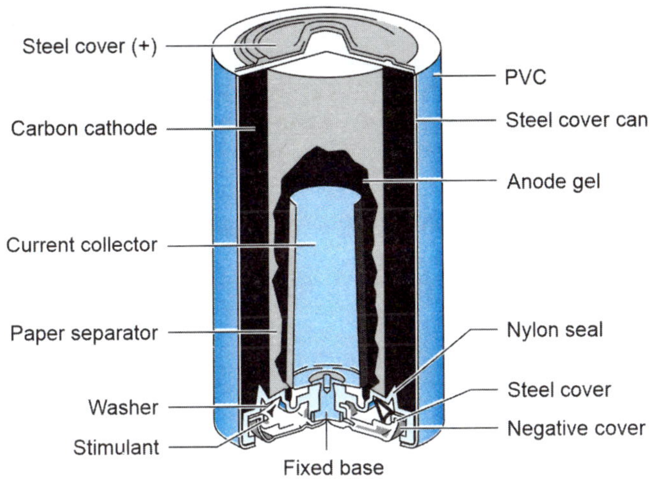

Fig. 4.5: Schematic diagram of alkaline batteries

The chemical reactions of alkaline battery are:

$$Zn_{(s)} + 2MnO_{2(s)} \rightleftharpoons Mn_2O_{3(s)} + ZnO_{(s)}(e^\circ = 1.43\ V)$$

i. Depending on the composition of the active materials of the plates, there are four types of battery. These are:
   a. Nickel iron (or Edison).
   b. Nickel- cadmium (or Nife).
   c. Silver zinc.
   d. Alkum battery.
ii. Depending on the method of assembling, these batteries are classified as sealed and non-sealed cells or battery.
iii. Depending on the design of plates, the alkaline battery is classified as an enclosed pocket and open pocket type battery.

Advantages of Alkaline Battery

- It has better dimensional stability.
- In continuous and intermittent applications were performed well with this battery.
- It has also low internal resistance.
- It has both low and high rate of discharge perform where it require.
- It has high energy density.
- It has longer durability and self-life.
- It performs is independent at ambient temperature as well as at low temperature.
- Low leakage

Disadvantage of Alkaline Battery

- Little costlier.

4.8 THERMOCOUPLE

It is also called thermal junction, thermoelectric thermometer, or thermal measuring device. It is used for measuring the temperature at one particular point with the help of sensor. The sensor that measures the temperature in the form of an **electric current** or **EMF**.

The two wires of different metals which are connected together at the ends are welded in thermocouple or junction. The junction is created in the welded portion and the temperature is measured at this thermoelectric couple. The variations in temperature of the wire induces the voltages.

Working principle of thermocouple: The thermocouple working principle depends on the three effects.

Seebeck effect: The seebeck effect is the phenomenon of thermocouple in which a temperature difference between two dissimilar electrical conductors produces a voltage difference. This voltage difference is observed when a closed circuit is formed by jointing two dissimilar metals at two junctions and junctions are maintained at different temperatures. An electromotive force (EMF) is induced in this closed circuit. The amount of induced EMF is vary for different metal combinations and is also proportional to the temperature difference of the junctions. This is the basic thermocouple working principle.

In short, *it is a phenomenon in which the temperature difference between the two different metals induces the potential differences between them.* The Seebeck effect produces small voltages for per Kelvin of temperature differences.

Peltier effect: On contrast, the Peltier effect is the inverse form of Seebeck effect. It states that *the temperature difference can be created between any two different conductors by applying the potential difference between them.*

Thompson effect: The Thompson effect state that *when two dissimilar metals join together and if they create two junctions then the voltage induces in the entire length of the conductor due to temperature gradient.* The physical term is the temperature gradient which shows the direction of current and rate of change of temperature at a particular location.

Construction of thermocouple: The thermocouple consists two different metals wires. These metal wires are connected together at the junction point. This junction considers as the measuring point at one area. The junction point is categorized into three types as shown in Fig. 4.6.

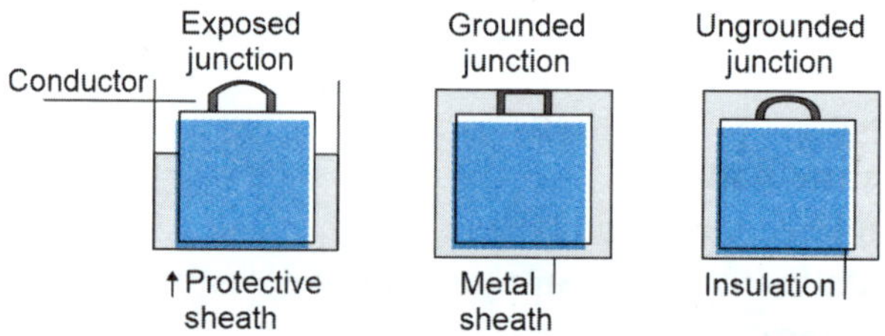

Fig. 4.6: Junction points of thermocouple

i. **Ungrounded junction:** In ungrounded junction, *the conductors are totally isolated from the protective sheath.* The junction also works in high pressure application. Main uses of this junction are to decrease the deviate magnetic field effect.

ii. **Grounded junction:** In such type of junction, *the metals and protective sheath are joined together.* The junction can measure the temperature in acidic atmosphere and reduces the noise effectively.

iii. **Exposed junction:** In such type of junction uses in the area where fast response requires. The exposed junction is used for measuring the temperature of the gas. The temperature sensor metal used in this junction varies depend on range of different temperature.

The thermocouple is made up of copper-nickel alloy and iron that forms sensing element of hot junction. This consist of two junctions, one is connected with transmitter for cold junction and other is associated in the process for hot junction. The thermocouple is shown in Fig. 4.7. In this picture, two different metals are couple together and form two junctions. The junctions with temperature of hot and cold is mentioned. When the temperature of the junctions varies, then the electromagnetic force generated in the circuit. For example, if temperature rises in hot junction of thermocouple produces a voltage which is greater that the cold junction. The difference between the two voltage is proportional to the difference in temperature and can be measure by using voltmeter in between junction.

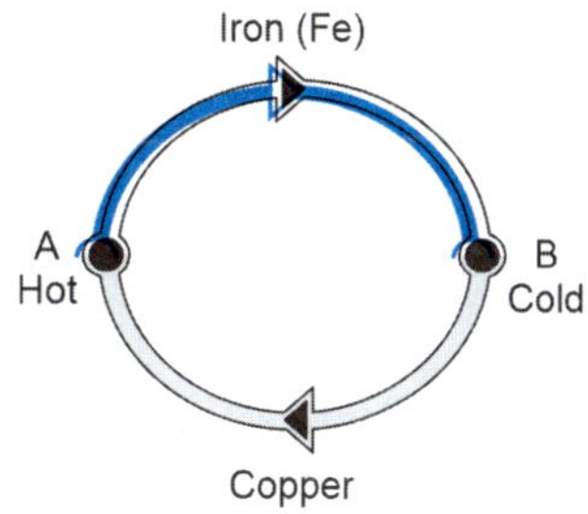

Fig. 4.7: Thermocouple

Advantages: The following are the advantages of the thermocouples.

i. Fast response time.
ii. Temperature range is wide.
iii. The thermocouple is cheap and best than the other temperature measuring devices.

Disadvantages

i. Ensure long life, need more care to protect against contamination.
ii. Inability in very high precision measurement because of very low accuracy.
iii. Located at a very large distance from the measuring device that causes more errors in the result.

4.9 ELECTROMAGNETIC INDUCTION

Electromagnetic induction is the phenomenon of production of EMF in a conductor by changing magnetic field or by change in magnetic flux around the conductor. When there is movement of magnet or coil. The magnetic flux linked through the coil changes. This causes an EMF which flows through the coil.

Michael Faraday who was an English Scientist contributed his research work on electro-chemistry and electromagnetism. His main discoveries on the principles of electromagnetic induction where he established the basic principles and the concept of electromagnetic theory in physics. The milestone of his contribution in the field of physics in the invention of electromagnetic rotary devices which formed the foundation of electric motor technology and his effort towards electricity became practical for use in technology.

4.10 FARADAY'S LAWS OF ELECTROMAGNETIC INDUCTION

It is the basic law of electromagnetism in which the magnetic field will interact with an electric circuit to produce an induced EMF. This phenomenon is known as electromagnetic induction and it is also called Faraday's law.

In 1831, he proposed the law of electromagnetic induction. It is the observation or experimental evidence of results conducted by Faraday. In this experiment, he explains the relationship between electric current and magnetic field. Based on his experimental results, he put forwarded two laws of electromagnetic induction that are Faraday's law of electromagnetic induction.

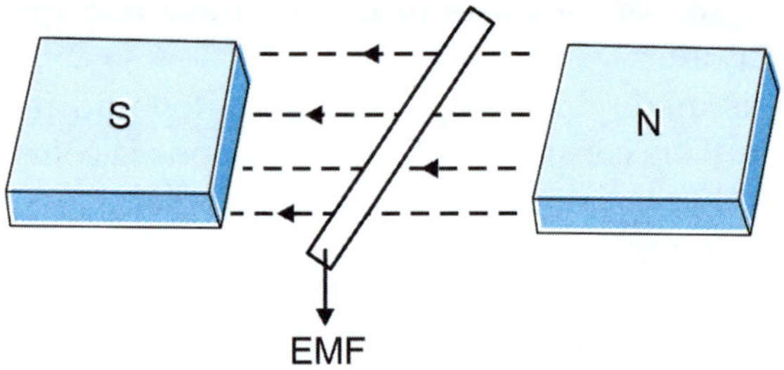

Fig. 4.8: Electromagnetic induction

4.10.1 Faraday's First Law

Whenever a conductor is placed in a varying magnetic field, an electromotive force is induced. If the conductor circuit is closed, a current is induced, which is called induced current.

Magnetic field can be changed as follows:

1. Changed by moving coil towards or outward from magnet field.
2. Changed by moving magnet towards or outward from the coil.
3. Changed by rotating the conductor coil relative to magnet.
4. Changed by area of coil relative to magnetic field.

4.10.2 Faraday's Second Law

Faraday's second law of electromagnetic induction states that *the induced EMF in a coil is equal to the rate of change of flux linkage.*

The magnetic flux is the product of number of turns in the coil and the flux associated with in. The formula of Faraday's law is given below:

$$\varepsilon = -N\Delta\phi/\Delta t$$

Where, ε = electromotive force,

Φ = magnetic flux, and

N = number of turns.

According to Faraday's law of electromagnetic induction, rate of change of flux linkages is equal to the induced EMF.

4.10.3 Induced EMF and Flux Relation

Faraday conducted a set of experiment where he confirmed the relationship between inducted and flux. They are:

In his first experiment, he confirmed that the current is induced in circuit only when the strength of magnetic field is varied. To confirm the relation galvanometer is connected in the wire where it shows deflection only when a magnet was moved towards or away from the wire.

In his second experiment, he confirmed that electromagnet is formed by passing a current through an iron rod. He also observed that when there is a relative motion between the coil and magnet, an EMF is generated. When the magnet is kept motionless then there is no deflection in the ammeter. The galvanometer shows deflection only the magnet is in motion or the coil in motion relatively.

In his third experiment, he kept the magnet at one position and he made the coil which carrying current in a relative motion. He observed that the galvanometer shows deflection only when the coil is in motion.

Applications of Faraday's Law

i. Transformers working on the principle of Faraday's law of electromagnetic induction.
ii. Induction cooker working on the principle of mutual induction which is the basis of Faraday's law.
iii. The main role of induced EMF in an electromagnetic flowmeter which recorded the velocity of the fluids is the applications in fluid mechanics.
iv. Faraday's law is also applicable in some musical instruments like electric guitar and electric violin.

4.11 LENZ'S LAW OF ELECTROMAGNETIC INDUCTION

Lenz's law states that *the direction of an induced EMF is such that it will always opposes change in magnetic flux that produced it*. In other words, an EMF induced in the circuit will always oppose the motion or change which induced EMF. This can be applied in the analysis of inductance.

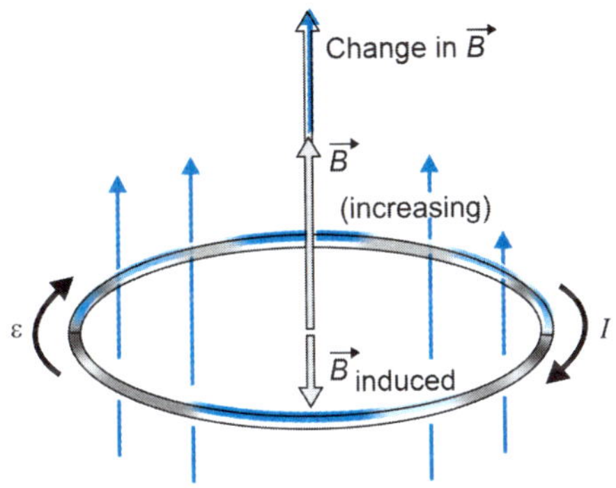

Fig. 4.9: Lenz's law

$$\varepsilon = -N\frac{\partial \Phi^{B}}{\partial t}$$

where,
ε = induced emf,
N = number of turns in coil (in figure $N = 1$)
$\partial \Phi^{B}$ = changing magnetic flux, and
∂t = rate of change in time

The negative sign indicates that the direction of an induced EMF and change in magnetic fields are in opposite direction.

4.12 FLEMING'S LEFT-HAND AND RIGHT-HAND RULE

In a magnetic field, when a current carrying conductor is placed a force is act on the conductor. The direction of the force can be acknowledged by using Fleming's left-hand rule. Similarly, when a current passed through the conductor and it is placed under a magnetic field, a force acts on the conductor. The direction of this force can be acknowledged by using Fleming's left-hand rule. Likewise, if a moving conductor is brought under a magnetic field, electric current will be induced in that conductor. The direction of the induced current can be found by using Fleming's right-hand rule. It is important to note that these rules do not determine the magnitude, instead show only the direction of the three parameters (magnetic field, current, force) when the direction of the other two parameters is known and third parameter can be depicted with Fleming's rules. Fleming's left-hand rule is mainly applicable to electric motors and Fleming's right-hand rule is mainly applicable to electric generators.

4.12.1 Fleming's Left-Hand Rule

Fleming's left-hand rule states that *when a current carrying conductor is placed in an external magnetic field, the conductor experiences a force perpendicular to the field and to the direction of the current flow.*

Keep the left hand three fingers in the position illustrated in Fig 4.10. These fingers represent the orthogonal axes. The thumb finger, middle finger and forefinger of left hand kept at right angel to each other, then the thumb finger points to the direction of force, the middle finger shows the direction of current and the forefinger points the direction of magnetic field.

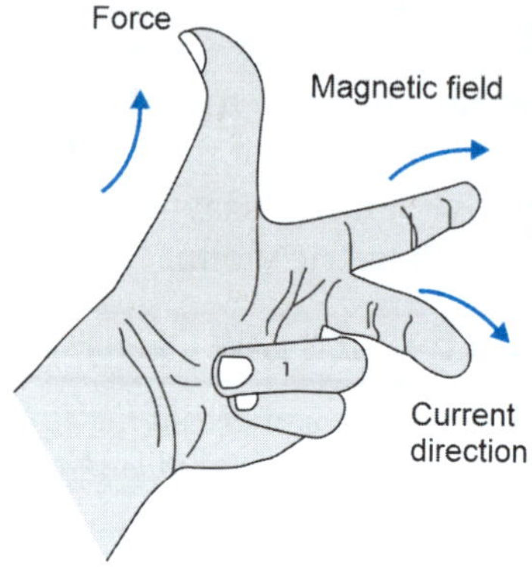

Fig. 4.10: Left-hand rule

Fleming's left-hand rule is mainly applicable to electric motors, according to this rule placing three fingers at right angel to each other that explains the force experienced in the electric motors. In a motor, if the current flows through the conductor which is in the direction of middle finger, the conductor is placed in the magnetic field which is in the direction of forefinger at right angel to each other. So, the force exerted by the electric motor which is in the direction of thump finger as shown in Fig. 4.11.

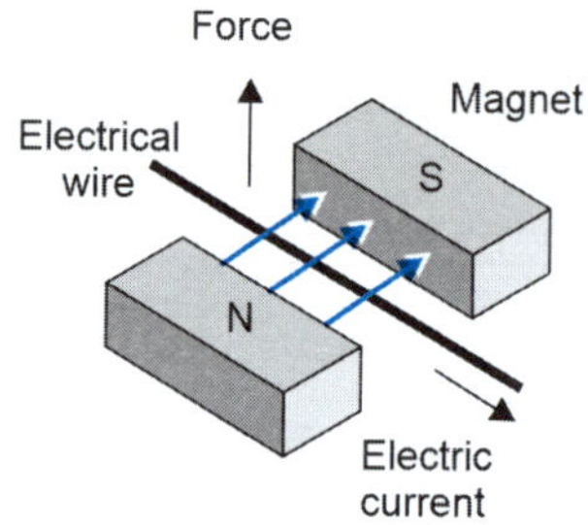

Fig. 4.11: Electric motor

4.12.2 Fleming's Right-Hand Rule

Fleming right-hand rule states that *when a conductor moves in a magnetic field, an electric current or EMF is induced in the conductor. If the conductor forcefully moved inside the magnetic field, there is a relationship between the direction of applied force, magnetic field and the current. It is determined by Fleming right-hand rule.*

According to this rule, stretch three fingers of right hand at right angle to each other. If the forefinger represents the direction of magnetic field of conductor, the thumb finger points the direction of motion or force act on conductor, then the middle finger points the direction of induced current.

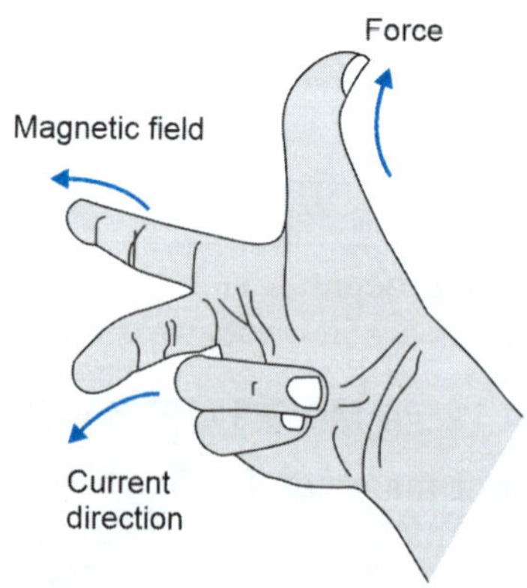

Fig. 4.12: Right-hand rule

The right hand thumb rule is also called Maxwell's corkscrew rule. Let us consider a driving a corkscrew in the direction of the current, then the direction of the corkscrew is in the direction of the magnetic field.

4.13 SELF INDUCTION

When a current flow in a coil, a magnetic field is set up in it (Fig. 4.13a). If the current through the coil is changed, the flux linked with the coil also changes. An induced EMF is set up in the coil. According to Lenz's law, the direction of induced EMF is such as to oppose the change in current. When the current is increasing, the induced EMF is against the current (Fig. 4.13b). When the current is decreasing, the induced EMF is in the direction of current (Fig. 4.13c). This phenomenon is called *self-induction* and the induced EMF is called *back EMF*.

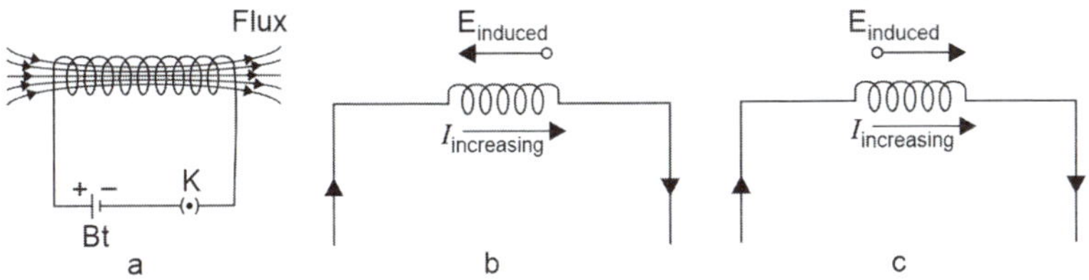

Fig. 4.13: Self induction: The circuit experiment

When the current is switched on in a coil, self induction opposes the growth of current. Hence the current in a coil increases slowly and takes some time to increase from zero to maximum value. During this time energy is absorbed. Similarly, when the current is switched off in a coil, self induction opposed the decay of current. So, the current in a coil does off instantaneously but takes some time (Fig. 4.14). When current is switched off the stored energy is delivered in the form of spark.

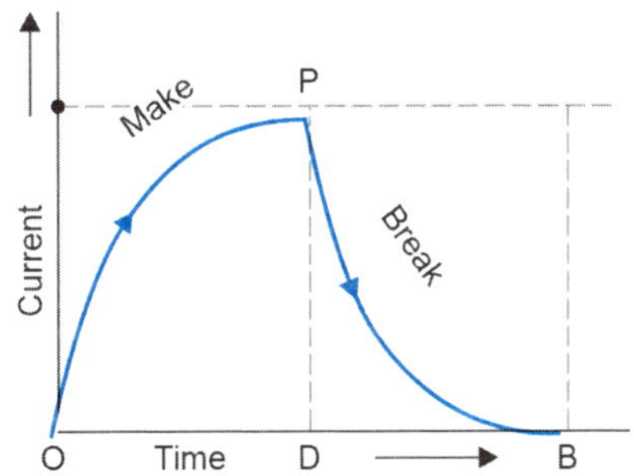

Fig. 4.14: Self induction graph

The phenomenon of the production of an induced EMF in a circuit itself due to the change in current through it is called self induction and the induced EMF is called back EMF.

Self inductance: The magnetic flux Φ produced in the coil is directly proportional to the current I in the coil, i.e.

$$\Phi \propto I$$

Or $$\Phi = LI \qquad ...(4.6)$$

Where, L is the proportionality constant called coefficient of self induction or self inductance of the coil.

From Eq (4.6), if I = 1 amp, then $L = \Phi$

∴ Self inductance of coil is the total magnetic flux that lined with the coil when 1 unit of current is passed through it.

According to Faraday's law, when the flux changes, the back EMF induced in the coil is given by,

$$\varepsilon = -\frac{d\Phi}{dt} \qquad ...(4.7)$$

Substitute the value of Φ in Eq. (4.7), we get

$$\varepsilon = -L\frac{dI}{dt}$$

$$\frac{dI}{dt} = 1 \text{ unit}$$

then, $$[\varepsilon] = L$$

Self inductance of a coil is numerically equal to the induced EMF when current in, it is changing at unit time.

SI unit of self-inductance is Henry (H).

From equations (4.6) and (4.7) we have,

One henry is the self inductance of coil of wire or circuit, if 1 amp current passed through the coil which produce a flux of 1 weber in it.

One henry is the self inductance of coil of wire or circuit, if 1 volt of induced EMF is produced in the coil is due the rate of change of current of 1 amp/sec.

4.14 MUTUAL INDUCTION

Consider two coil of wire say primary coil (P) and secondary coil (S) which is placed very close together as show in Fig. 4.15. When current is applied to P changes, the magnetic flux linked to the coil S also changes. Hence an induced EMF is set up in the coil S. This phenomenon is called *mutual induction*. If any circuit which is mutually inducted is called coupled circuit.

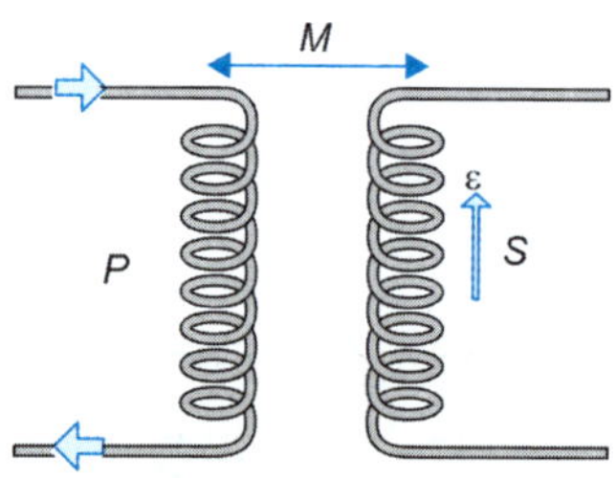

Fig. 4.15: Mutual induction

Let us consider I is the current flowing to the primary coil P. So that the magnetic flux Φ inducted in secondary coil is depends on I.

$$\Phi \alpha I$$

Or $$\Phi = MI \qquad \text{...(4.8)}$$

Where M is the cofficient of mutual induction or mutual inductance of the two circuits. If $I = 1$, then $\Phi = M$.

From the above equation, the mutual inductance of two circuit is equal to amount of magnetic flux linked with one circuit only when 1 unit of current flows through the other circuit.

The EMF induced in S is given by

$$\varepsilon = -\frac{d\Phi}{dt}$$

$$\varepsilon = -M\frac{dI}{dt} \qquad \text{...(4.9)}$$

If $\frac{dI}{dt} = 1$ unit, then $$[\varepsilon] = M$$

The mutual inductance of two circuit is equal to the induced EMF in one circuit when the rate of change of current in the other circuit in 1 unit.

In Eq. (4.9), if ε is induced EMF in volt and $\frac{dI}{dt}$ in unit ampere current per second, then the mutual inductance is one henry. Hence, *henry is defind as mutual inductance between two circuit is one unit and if a change of current is 1 amp per second in one circuit which induces 1 volt of EMF in other circuit*. This phenomenon is called mutual induction.

4.15 EDDY CURRENTS

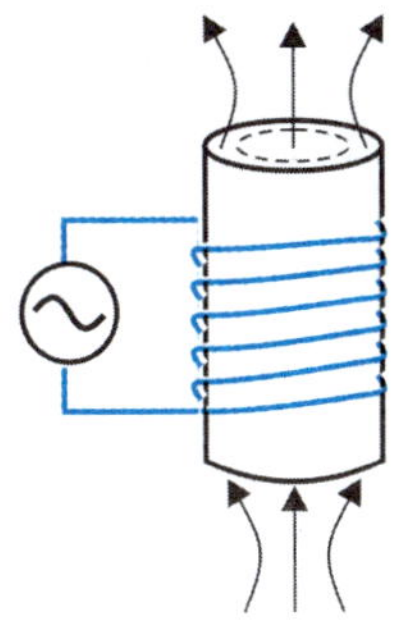

Fig. 4.16: Eddy currents

According to Faraday's law of electromagnetic induction, a loops of electrical current are induced within conductors by changing magnetic field in the conductor. Eddy currents is in a plane at perpendicular to the magnetic fields of conductor.

Let us consider the coil of wires which are surrounded with a metal core as shown in Fig. 4.16. When a current passed through the coil that produce a oscillating flux through the metal as shown in Fig. 4.16 with dotted line. Changing flux in that line induces EMF around the line, this is similar to other EMF induced in core and forms circular current within the coil. This induced circular current is called eddy currents, flow in metal objects subjected to a changing magnetic field.

Eddy currents are named because of the appearance of current like eddies or swirls. When a conductor is subjected to the changing magnetic field, the induced current in the conductor is termed as eddy currents.

The applications of eddy current can seen in the speedometer, a small bar magnet is connected to the main shaft of the vehicle. According to the speed of the vehicle, it rotates. Due to the effects of the eddy currents that form by the motion of the rotating which is opposed and the pointer gets deflected through a certain angle. The pointer attached to the calibrated scale indicates the speed of the vehicle.

4.16 INDUCTORS

An inductor is a passive electrical component which means consumes energy and not produce it. It consisting of a coil of wire where current passing through it and that causes a relationship between magnetism and electricity. It can store the energy in a magnetic when current flows through it and consists of insulated wire wound into a coil. Inductors are also called coils or chokes. 'L' is the electrical symbol for an inductor.

Symbol of Inductor (Fig.4.17)

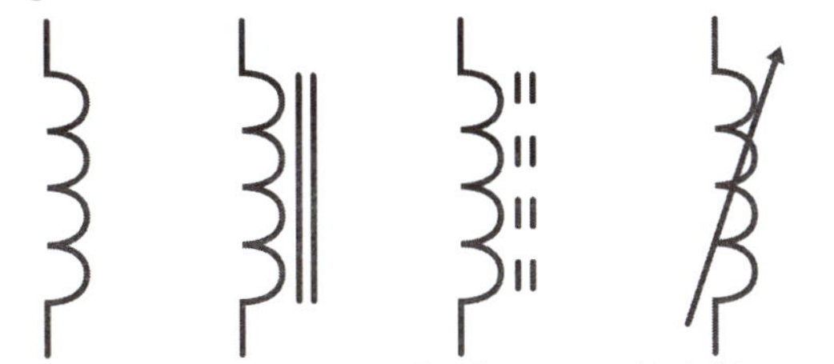

Fig. 4.17: Inductor symbol

4.17 TRANSFORMER

It is an electrical device can be used for changing the level of voltage between the circuits which means converting a low alternating voltage into high alternating voltage and vice versa. As it is control voltage device as well working under the principle of mutual induction between two coils.

Construction: A transformer consist of two coils namely primary (P) and secondary (S) coil. They are insulated from each other and wound on a soft iron core that are laminated. The core is joined in the form of strips with high mutual inductance (M). When current pass through the coil P, it creates a varying magnetic flux. According to Faraday's law of electromagnetic induction, this induces an EMF in the coil S which is linked to the core having a coil P. This is mutual induction.

Transformers are widely used in safety reasons, for example, in construction sites when using power tools and in domestic bathroom situations in shaver units. They are also used in doorbell operation and to power electronic equipment, battery chargers, televisions, computers, alarm systems, etc. Transformers will vary substantially in construction, also size and shape depending on their application.

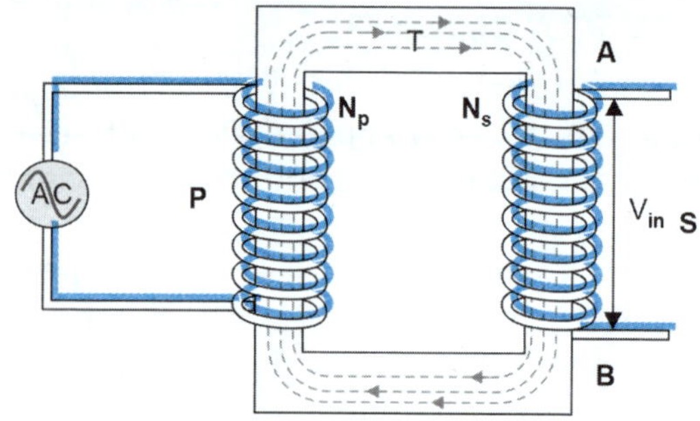

Fig. 4.18: Transformer

Transformer Symbols (Fig. 4.19)

Fig. 4.19: Schematic symbols of transformer

4.18 TYPES OF TRANSFORMER

There are different types of transformers are in use in the electrical power system like generation, distribution, transmission, utilization of electrical power, etc.

Transformers are classified based on the following factors:

i. Working voltage range.
ii. The medium used in the core.
iii. Winding arrangement.
iv. Installation location.

4.18.1 Based on Voltage Levels

This type of transformer which is commonly used in this field and can be transformed voltage in between the coil of wire, so depending upon voltage it is classified as:

Step-up transformer: In this type of transformer having two coils, namely primary and secondary, where the secondary output voltage will be higher than input. It will be used in power generator and the power grid.

Step-down transformer: In this type of transformer, secondary output voltage will be lower than primary input voltage. It can be lowered the voltage level where it required in electronic devices and suitable for electronic supply.

4.18.1.1 Step-down transformer

Transformer is a static apparatus with no moving parts. It transforms electrical power from primary (P) to secondary (S) in voltage and current. But there is no change in frequency. Because the same magnetic flux lines cut both coils of a transformer and transformed it. The induced EMF in the secondary winding is proportional to the number of turns on both the primary and secondary windings.

If the number of turns on the secondary winding (N_S) is less than the number of turns on the primary winding (N_P), then the secondary output voltage will be less than the primary input voltage. This type of transformer is called a step-down transformer which is illustrated in Fig. 4.20.

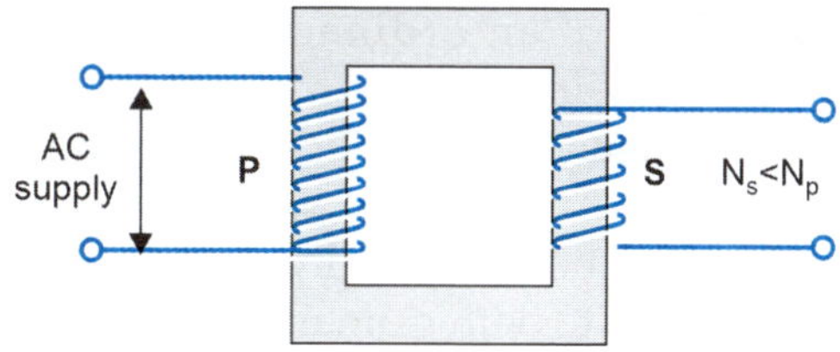

Fig. 4.20: Step-down transformer

4.18.1.2 Step-up transformer

If the number of turns on the secondary winding (N_S) of a transformer is greater than the number of turns on the primary winding (N_P), then the secondary output voltage will be greater than the primary input voltage. This type of transformer is called a step-up transformer which is illustrated in Fig. 4.21.

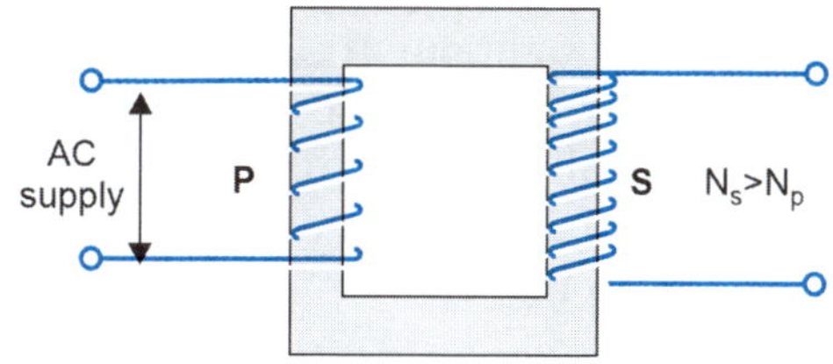

Fig. 4.21: Step-up transformer

4.18.2 Based on the Medium of Core Used

This type of transformer can be classified based on the type of core used in it.

Air core transformer: In this type of transformer, no physical core is used. So it is called air core transformer in which the flux linkage between primary (P) and secondary (S) winding is only through the air. The coil or windings wound on the non-magnetic strip (Fig. 4.15).

Iron core transformer: In this type of transformer used an iron core for windings. It is wound on multiple iron plates stacked together, which provides a perfect linkage path to generate flux (Fig.4.18).

4.18.3 Based on the Winding Arrangement

In this type of transformer are classified based on winding of coil wound over a laminated core.

Autotransformer: It is a type of transformer which has only one winding. The name "auto" in prefix refers to the single coil acting alone (it means "self") and is not of any automatic mechanism. It is working in similar way of two winding transformer but varies in the way the primary (P) and secondary (S) windings of the transformer that are interconnected in induction purposes.

One single winding only used in autotransformer. This winding is used as primary winding as well as secondary winding. But in case of two windings transformer, it has two different windings and one is used for primary and another is for secondary. A circuit diagram of autotransformer is shown in Fig. 4.22.

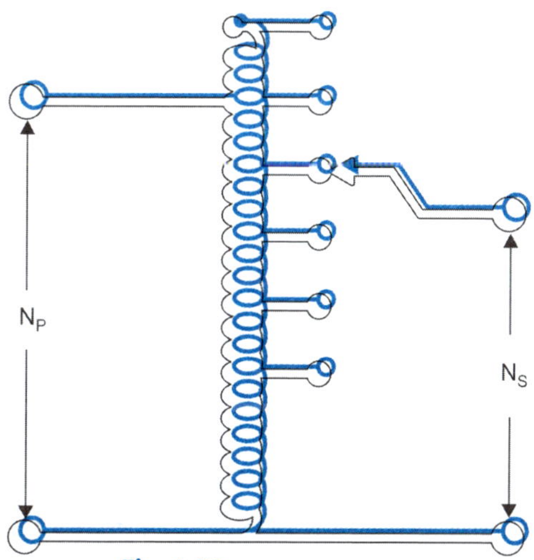

Fig. 4.22: Autotransformer

In case of an ordinary transformer, P and S are electrically insulated from each other but it is linked by flux density magnetically as shown in Fig. 4.23. In autotransformer, P and S are connected electrically as well magnetically. Actually, a part of the lengthy winding is common to both primary (P) and secondary (S).

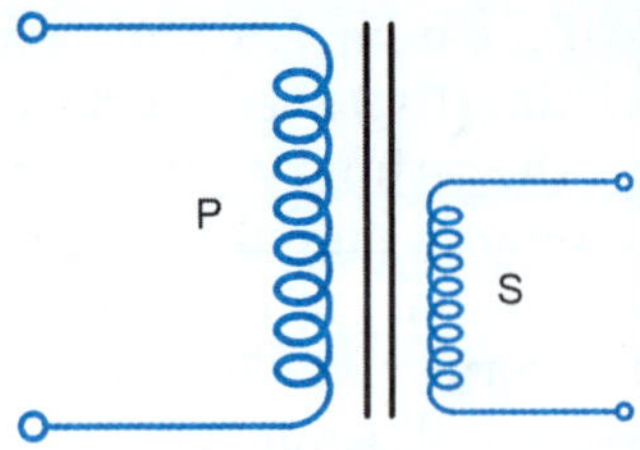

Fig. 4.23: Ordinary transformer

Based on construction it has two types. In type 1 construction, there is continuous lengthy winding where primary can be adjusted to fix number turns in primary and it is brought out at convenient points determined by the desired secondary voltage. However, in another type of autotransformer, there are two or more distinct coils which are electrically connected to form a continuous winding. The construction of autotransformer is shown in Fig. 4.24.

Let us consider the primary winding AB in which a variable point is consider at C, such that CB acts as a secondary winding. The supply voltage is applied across AB which is variable by number of turns and the load is connected across CB. The point C may be fixed or variable. When an AC voltage V_1 is applied across AB, an alternating flux is set up in the core, as a result, an EMF E_1 is induced in the winding AB. A part of this induced EMF is taken in the secondary circuit.

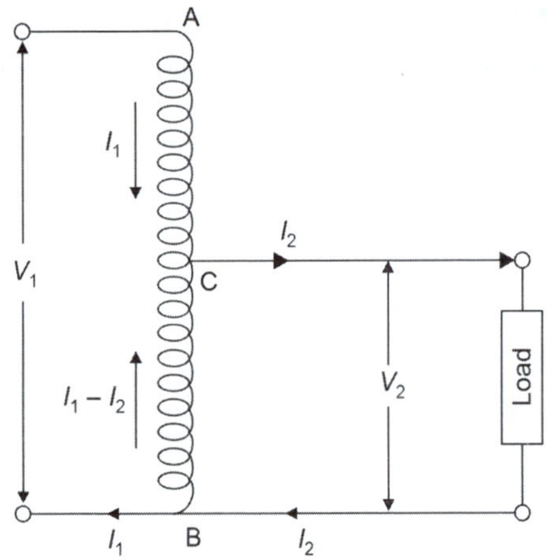

Fig. 4.24: Construction of autotransformer

If , V_1 = primary voltage

V_2 = secondary voltage

I_1 = primary current

I_2 = load current

N_1 = number of turns between A and B

N_2 = number of turns between C and B

Neglecting no-load current, leakage reactance and losses,

therefore, $V_1 = E_1$ and $V_2 = E_2$

Transformation ratio is given by:

$$K = \frac{V_2}{V_1} = \frac{N_2}{N_1} = \frac{I_1}{I_2}$$

Limitations: An autotransformer does not provide electrical isolation between its windings as an ordinary transformer does. A failure of the insulation of the windings of an autotransformer can result in full input voltage applied to the output. If there is a break in the part of the winding, then the transformer acts as an inductor in series with the load.

4.18.4 Based on Usage and Install Location

According to the need and location, it is classified as the power transformer, distribution transformer measuring transformer, and protection transformer.

4.18.4.1 Power Transformer

The size of power transformers is big and it is suitable only for high voltage (>33 KV) power transform applications. So it will very useful in case of power generation stations and transmission substation. It has high insulation level.

4.18.4.2 Distribution transformer

The generated power from the power plant can be distributed to remote areas by using special type of transformer is called *distribution transformer* as shown in Fig. 4.25. Basically, it is used for the distribution of electrical energy at low voltage is less than 33 KV in industrial purpose and 440 V–220 V in domestic purpose.

- It works at low efficiency at 50–70%
- Small size
- Easy installation
- Low magnetic losses
- It is not always fully loaded

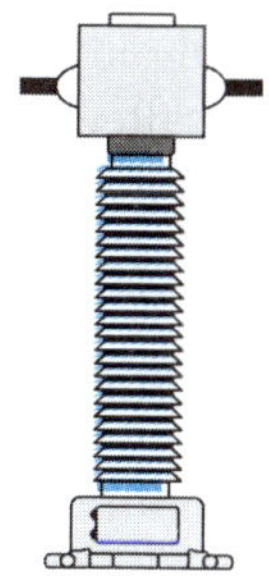

Fig. 4.25: Distribution transformer

4.18.4.3 Measurement transformer

Used to measure the electrical quantity like voltage, current, power, etc. These are classified as potential transformers, current transformers, etc. This type of transformer should have a characteristic like the accurate measuring of currents and voltages with high insulation and negligible insertion loss. It is also widely used on instrumentation and equipment for photovoltaic, wind energy, automation and industrial.

4.18.4.4 Protection transformers

This type of transformers can be used in component protection purpose. It will vary with its accuracy when compared with measuring transformers, that is, major difference and protection transformers is the accuracy that means that the protection transformers should be accurate as compared to measuring transformers.

4.18.5 Based on the place of use

Transformers are classified based on the places where they are used. The transformer covered with proper roof is called indoor transformer, whereas in outdoor transformers perform like distribution type transformers.

4.18.6 Losses in a transformer

Losses in a transformer can be determined by its efficiency and should have energy losses, which means the power supplied to the transformer's input terminal should be equivalent to the power output received from the transformer. So that in case of ideal transformer should have output power that is equal to input power with zero energy losses. But practically, input and output powers of the transformer will not equal due to energy losses. It is a stationary device because it does not have any movable physical parts inside, so we cannot observe mechanical losses but we can observe electrical losses.

4.18.7 Types of Loses in a Transformer

Different types of losses are discussed below.

4.18.7.1 Iron Losses

This type of iron losses occurs through flux changes within the transformer's core and it is also called core loss due to loss in its core of transformer. Magnetic properties of core material play a major role in this type of loss and the core material is also made up of iron, so it is called iron losses. This type of loss can be categorized into two types: i. Hysteresis loss and ii. Eddy current loss.

i. Hysteresis loss

This type of hysteresis loss where the alternating current is given to the core of the transformer then reverse magnetic field will be observed. Core material of transformer plays main role in this type of loss and the motion of the magnetic field will also one of the reasons for the loss. The high-grade core material can be used to reduce this loss. This loss can be represented by Steinmetz's formula and hysteresis loss is calculated by,

$$P_h = k_h f b_m x \text{ watts}$$

Where,

k_h = constant that depends on the quality and volume of the core material

b_m = highest flux density in the core

f = alternatintg flux frequency

x = constant of the Steinmetz (1.5 to 2.5)

ii. Eddy current loss

When the current passed through the coil that produce the flux. This flux is linked with a closed circuit, this causes an induced EMF in the circuit and the current flows in the linked coil. The flow of the current depends upon the sum of EMF in the circuit and the resistance offered by the circuit. As we know the core is made of conducting material, this induced EMF current flows within the body of the material. These circulating currents are called eddy currents. This type of losses occurs during a changing magnetic field in the conductor. As is not responsible for doing any useful work and it produces a loss (I^2R loss) in the magnetic material known as an eddy current loss and it can be reduced by laminating the core. Lamination on the core can reduced the area of each section and inducing EMF also reduced. So that current is passing through it also reduced and resistance of eddy current path increases. The eddy current loss can be represented by:

$$P_e = k_e b_m^2 t^2 f^2 V \text{ watts}$$

Where,

k_e = Co-efficient of eddy current which depends on the magnetic material

b_m = maximum value of flux density in wb/m^2

t = thickness of lamination in meters

f = frequency of reversal of the magnetic field in Hz

V = volume of magnetic material in m^3

4.18.7.2 Copper Loss or Ohmic Loss

This will occur due to ohmic resistance in the windings of the transformer. Ohmic property of the material is the main reason for the loss. Ohmic loss is nothing but the voltage drops across the circuit which is directly proportional to current. The primary and secondary windings of the transformer are having current I_1 and I_2, then the resistance of these windings is R_1 and R_2. So the copper losses that occurred in the windings are $I_1^2R_1$ and $I_2^2R_2$ respectively. So, the entire copper loss will be

$$P_c = I_1^2R_1 + I_2^2R_2$$

These losses (*generally called as power loss* $P = IV = I^2R$ (*Ohm law*, $V = IR$)) also called variable or ohmic losses because these losses will change based on the load.

4.18.7.3 Stray Loss

It is the type of losses in a metallic region of transformer where magnetic flux leakage occurs. When compared with copper and iron losses, the percentage of stray losses is negligible. It is also called transformer tank, winding, etc. The less leakage current in the transformer, stray become negligible. So, if we control leakage of current, automatically stary loss reduced.

4.18.7.4 Dielectric Loss

It is the type of losses occurred due to insulating material property like quality of oil material and weakening of insulation material. So, it can reduce the efficiency of transformer and this causes in increasing dielectric loss. It rarely occurs as compared to the core and copper losses.

4.19 CAPACITORS

The capacitor is a type of component with two terminal which has the ability or "capacity" to store energy in the form of an electrical charge with a potential difference (static voltage) across its plates, almost like a small rechargeable battery. It has two electrical conductors that are detached with a material, called dielectrics placed in between capacitor plates in a small distance. The space may be occupied by vacuum or dielectric material. The ability to the store the energy in the form of charge is known as capacitance.

Construction and properties of capacitor: In general, capacitors have two or more parallel conductive plates which are not allowed to touch or connect each other. That means it is electrically separated either by air or by some form of a good insulating materials like waxed paper, mica, ceramic, plastic, or some form of a liquid gel. This material is commonly called dielectric materials.

DC current in the circuit cannot flow through the capacitor because of this insulating layer, as it blocks it allowing the electrical charge to accumulate near the plate.

Depending upon its application, voltage rating and usage, the metal plates of a capacitor can be square, circular or rectangular, or they can be of a cylindrical or spherical shape.

When an AC supply is given to the capacitor, the flow of current appears in the circuit and capacitor will not resist the current and it could have little flow. But the capacitor is connected to DC supply, a capacitor charges up to its supply voltage but blocks the flow of current through it because the dielectric of a capacitor is non-conductive and basically an insulator.

There are two types of electrical charge, namely positive charge is called protons and negative charge is called electrons. When a DC supply is given across a capacitor, the protons quickly accumulate on one plate, whereas electrons accumulate on the other plate. For every particle of +*ve* charge that arrives at one plate a charge of the same sign will depart from the –*ve* plate as shown in Fig. 4.26. So that the plates remain charge neutral and a potential difference due to this charge is established between the two plates. Once the capacitor reaches its steady state condition an electrical current is unable to flow through the capacitor itself and around the circuit due to the insulating properties of the dielectric used to separate the plates.

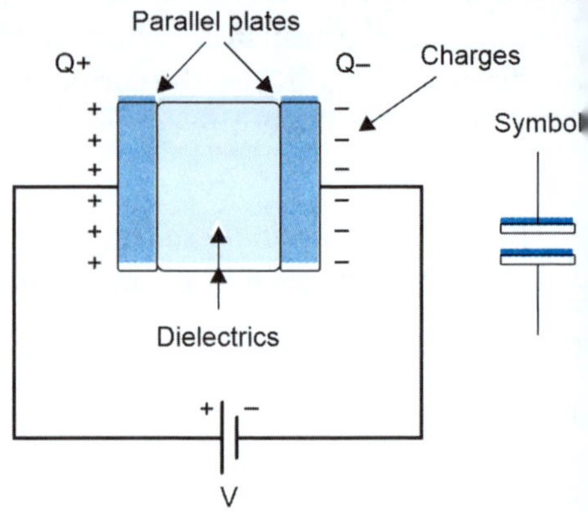

Fig. 4.26: Construction of capacitor

Capacitor charging current is nothing but the flow of electrons on the plate which continues to flow till the potential across the plate will be equal to the applied voltage. Once it reaches this situation called fully charge capacitor. The capacitor charge depends on the applying potential difference and when it is connected to voltage source, charges are starting to deposit on the plates.

The simplest form of capacitor is made up of parallel plates. It is made up of two metal plates at a distance parallel to each other, with its capacitance value in Farads, being fixed by the surface area of the conductive plates and the distance of separation between them.

A capacitor is applying with a voltage V and measuring the charge on the plates. The ratio of the charge 'Q' to the voltage 'V' will give the capacitance value of the capacitor.

Therefore, $$C = \frac{Q}{V} \quad ...(4.10)$$

Then, the quantity of charge on the plates as:

$$Q = C \times V \quad ...(4.11)$$

From Eq. (4.11), the magnitude of the charge Q stored on either plate of a capacitor is directly proportional to the potential difference V between the plates. Where C is a constant of proportionality called the capacitance of the capacitor.

The SI unit of a capacitance is the farad (F). 1 Farad = 1 coulomb/volt

4.19.1 Capacitance of a Capacitor

Capacitance is the electrical property of a capacitor and is the measure of a capacitors ability to store an electrical charge onto its two plates with the unit of capacitance being the **Farad** (abbreviated to F) named after the British Physicist—Michael Faraday.

It is defined as a capacitor has the capacitance of **one farad** when a potential difference of one volt is maintain between to plates to store a charge of **one coulomb**. Capacitance C is always positive and does not have negative units. However, the farad is a very large unit of measurement to use on its own sub-multiples of the farad are generally used such as microfarads, nanofarads and picofarads, for example.

Standard units of capacitance

i. Microfarad (μF) $1\ \mu F = \frac{1}{1{,}000{,}000} = 10^{-6}\ F = 0.000001$

ii. Nanofarad (*nF*) $1\, nF = \dfrac{1}{1{,}000{,}000{,}000} = 10^{-9}\, F = 0.000000001$

iii. Picofarad (pF) $1\, pF = \dfrac{1}{1{,}000{,}000{,}000{,}000} = 100^{-12}\, F = 0.000000000001$

Capacitance of capacitor is directly proportional to the area A of the plate which is in metre^2 and inversely proportional to '*d*' which is the dielectric thickness in meters.

In general, the equation for the capacitance of capacitor is given as:

$$C = \varepsilon\left(\frac{A}{d}\right) \qquad ...(4.12)$$

Where, ε is the permittivity of the dielectric material, ε_0 is the permittivity of a vacuum or "permittivity of free space" and has the value of the constant 8.84×10^{-12} farads per meter. It can also be written as:

$\varepsilon_0 = \dfrac{1}{(4\pi \times 9 \times 10^9)}$, or picofarads (pF) per meter as the constant, which is equal to 8.84 *pF* and it is the value of permittivity of free space.

4.19.2 Multi-plate Capacitor

The construction of multi-plate capacitor is shown below. Let us consider five plates that are connected to one terminal '*A*' and four plates are connected to another terminal '*B*'. All these four plates are connected and covered with dielectric on both sides. The five plates are placed in between terminal A and terminal B which is covered with dielectrics except outer portion of *A*. Therefore, this constitute multi-plate capacitor and its capacitance is given by

$$C = \frac{\varepsilon_0\varepsilon_r(n-1)A}{d} = \frac{\varepsilon_0\varepsilon_r(9-1)A}{d} = \frac{\varepsilon_0\varepsilon_r 8A}{d}$$

Nowadays, capacitors are classified based on characteristics and properties of their insulating material:

i. **Material 1:** Low loss rate with high stability, e.g. Mica, low dielectric ceramic, polystyrene.

ii. **Material 2:** Medium loss with medium stability, e.g. paper, plastic film, high dielectric ceramic.

iii. **Material 3:** Polarized capacitors with oxide layer as insultator, e.g. tantalum electrolytics.

4.19.3 Parallel Plate Capacitor

Let us consider a two parallel metal plates with area of each plate is *A* and separated by a distance *d*. The air is the medium between two plates. A charge +*q* is given to a plate 'P' and induces equalent negative charge −*q* on the surface of Q as shown in Fig. 4.27. According to the arrangement of plates, the force of electric lines created by the +*q* charge from plate P that reaches to the plate Q. This line of force are parallel to each other and perpendicular to the plates.

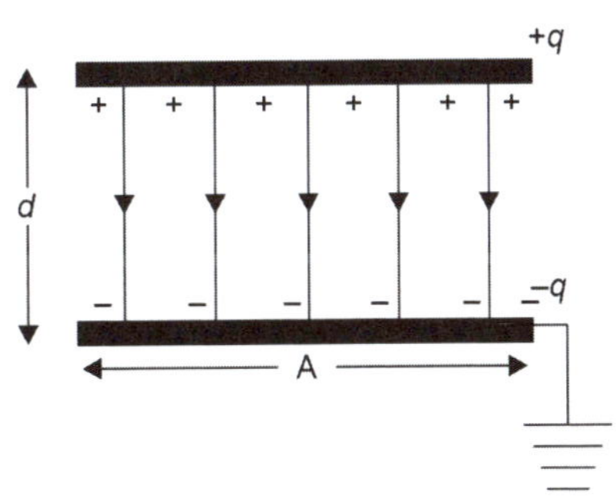

Fig. 4.27: Parallel plate capacitor

According to Gauss's law,

Electric field at a point between two plates $E = \frac{\sigma}{\varepsilon_0}$

Where σ = charge density in plate surface $= \frac{q}{A}$

The potential difference between the plates P and Q is

$$V = \int_d^0 -Edr = \int_d^0 -\frac{\sigma}{\varepsilon_0}dr = \frac{\sigma d}{\varepsilon_0}$$

The capacitance of parallel plate capacitor is

$$C = \frac{q}{V} = \frac{\sigma A}{(\sigma d/\varepsilon_0)} = \frac{\varepsilon_0 A}{d}$$

$$C = \frac{\varepsilon_0 A}{d}$$

4.19.4 Current-Voltage (I–V) Characteristic Curve

An I–V curve is also called current-voltage characteristic curve and it is a graphical representation of the relationship between applied voltage to an electrical device and the current flowing through that device. It can be used to determine the functions of electrical device in the circuit. Key properties of electronic devices can also be extracted from the shape and details of the curve, enabling greater insight into their operation.

Based on different types of electronic devices and their shapes, I–V curve also have many types.

Measuring and analyzing an I–V curve: By applying a series of voltage to the device an I–V curve is measured and analyzed. The current flowing through the device is measured based on each supplying. If a voltmeter is connected in parallel to the circuit, the supplied voltage to the circuit can be measured. But in case of measuring the current, an ammeter is connected in series in circuit as shown in Fig. 4.28.

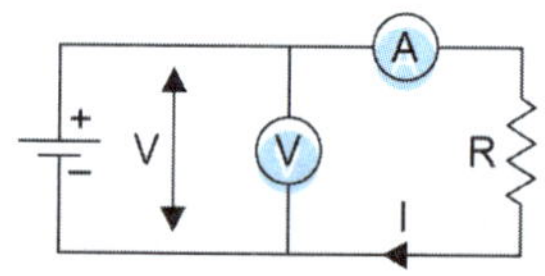

Fig. 4.28: An ideal resistor

Quadrants of an I–V curve: If the device can measure the source, then the device capable of measuring current with high accuracy and also supplying voltage.

The voltage used in an I–V measurement is mainly depending upon the device which is to be tested. For example, a solar cell may be tested between –1 V and 1 V, whereas LED may be tested with in higher range of 0 V to 10 V. So, the supplied voltage may vary depending upon the device that kept for testing. In some other cases, the applied voltage changes the characteristic of the electronic device. This will change the current in the circuit, even the voltage is kept constant. In that situation, the device should have a pause between setting a voltage and measuring the current.

The basic operation of an electronic device can be assumed from the position of the curve on the I–V graph. The graph can be split into four portions (i.e. I, II, III and IV) in the I–V characteristic curve as shown in Fig. 4.29. During the testing of device and the position in I–V curve reveals whether it is an active or passive device.

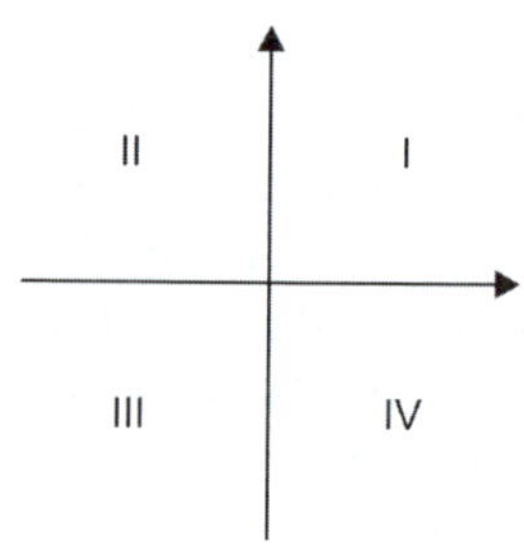

Fig. 4.29: I-V characteristics

A passive device I–V curve is only in I and III portion is given in Fig. 4.30. With same polarity (i.e. are both positive or both negative). Such type of devices only use the electrical power of the circuit. Similarly in active device, I–V curve is only in II and IV portion is given in Fig. 4.30 with opposite polarities. Such type of device creates electrical power. With the following example, it will be explained clearly.

Resistor I–V curve: Figure 4.30 shows a simplest I–V curve of an electronic is devices is called a resistor. It is a straight line which intercepts the origin and passes through protion I and III—making a resistor a passive device. According to Ohm's law ($V=IR$), the current at each voltage is proportional to the resistance. Therefore, the gradient of the line is equal to $\frac{1}{R}$, enabling the resistance to be extracted from the I–V curve.

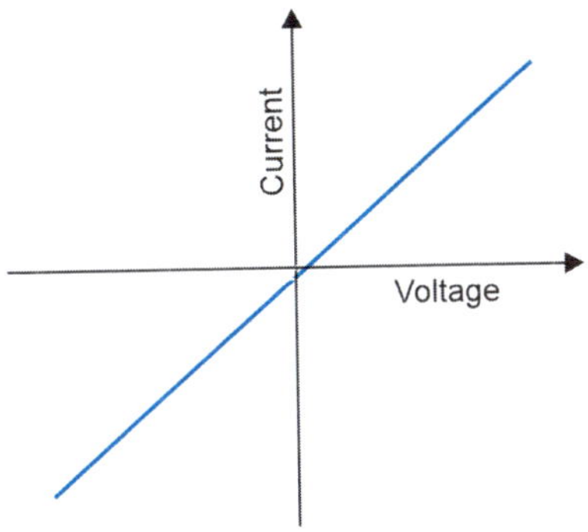

Fig. 4.30: Resistor I–V curve

Diode: Figure 4.31 shows a diode is an electronic device which allows current to flow only one direction. This can be seen in the I–V curve. At positive voltages, the curve rises substantially which indicates the current is free to flow in the device. At negative voltages, the current remains nearly at zero. At particular point there is a sufficient level of negative voltage (known as the 'breakdown voltage') will cause the diode to become conductive to negative current. Similar to a resistor, a standard diode is a passive device, operating only within portion I and III.

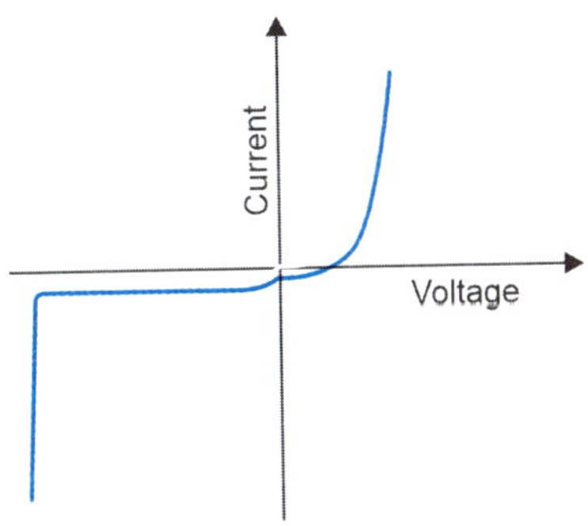

Fig. 4.31: Diode

Solar cell: Figure 4.32 shows a solar cell is a device that produce electricity with the help of sunlight and it converts it into electricity. In the dark, the solar cell simply acts as a diode. In the light, the photocurrent can be thought as a current source. However, when illuminated, the I–V curve shifts downwards into portion IV. In this situation, a solar cell behaves an active device. When the device is operating in this portion, it supplies electrical power to the circuit.

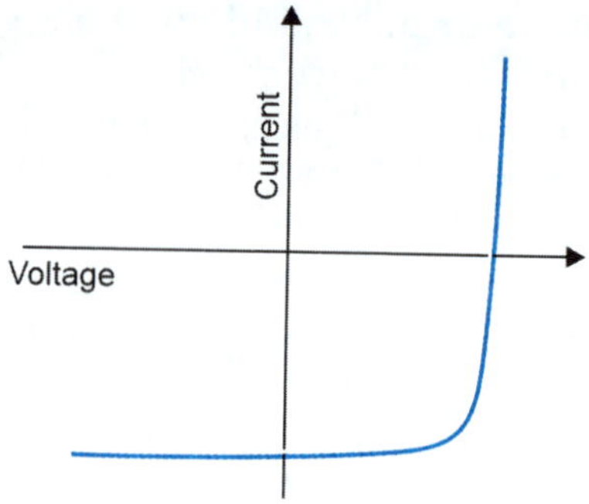

Fig. 4.32: Solar cell

4.19.5 The Farad

In the field of physics, capacitance is generally defined as the ability to hold an electric charge and the unit is farad.

The capacitance of a capacitor is one farad, then one coulomb of charge is store when a potential difference is maintain between the plates by one volt.

Alternately, the farad can be defined as *the capacitance that stores a charge of one coulomb across a voltage (or electric potential difference) having a magnitude of one volt*. The SI unit of electrical capacitance is named after the English scientist Michael Faraday. The SI base units of the farad are $s^4 \cdot A^2 \cdot m^{-2} \cdot kg^{-1}$. It can be noted that the values of most capacitors are expressed in farads, microfarads, and nano-farads.

4.19.6 Capacitors in Parallel and Series

Capacitor in series: If the capacitors are connected in series, the charging current (i_c) is given to the capacitor should be same for all capacitors that are connected series because it has only one path to follow. So that capacitors in series, all have the same current flowing through them as $i_T = i_1 = i_2 = i_3$ etc. As current flow to the capacitor is same, it will store the same amount of electrical charge, Q on its plates regardless of its capacitance. This is because the charge stored by a plate of any one capacitor must receive charge from the plate of its adjacent capacitor. Therefore, capacitors connected in series must have the same charge. $Q_T = Q_1 = Q_2 = Q_3$ ..., etc.

Construction: Let us consider the following circuit in which three capacitors are connected in series, namely C_1, C_2 and C_3. These capacitors are connected together with a supply voltage between points P and Q as shown in Fig. 4.33.

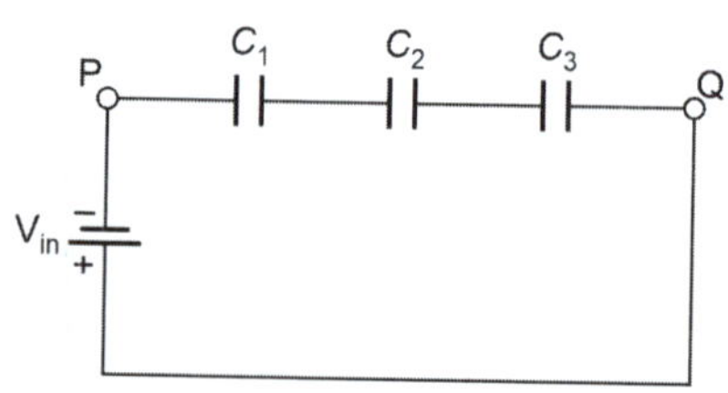

Fig. 4.33: Construction

The total capacitance of a capacitor which is connected in series circuit is obtained by adding up the reciprocals (1/C) of the capacitance values of the individual capacitors and then taking the reciprocal of the total, i.e.

$$\frac{1}{C} = \frac{1}{C_1} + \frac{1}{C_2} - \frac{1}{C_3} - \frac{1}{C_4} - \ldots - \frac{1}{C_n}$$

For example, if three capacitors are connected in series. Then the total capacitance of the circuit is given by:

$$\frac{1}{C} = \frac{1}{C_1} + \frac{1}{C_2} + \frac{1}{C_3}$$

$$C = \frac{1}{C}$$

Capacitor in parallel: If the capacitors are connected in parallel which means they are connected side by side in different path in an electronic circuit so that the same charge or current will not flow through each capacitor. When a input voltage (Vin) is applied to capacitors in the parallel circuit, each of it gets the different charges as shown in Fig. 4.34. The capacitor with high capacitance can receives more charge whereas the capacitor with less capacitance can receive less charge. For example, the capacitor of 10F will get more charge than the capacitor 5F.

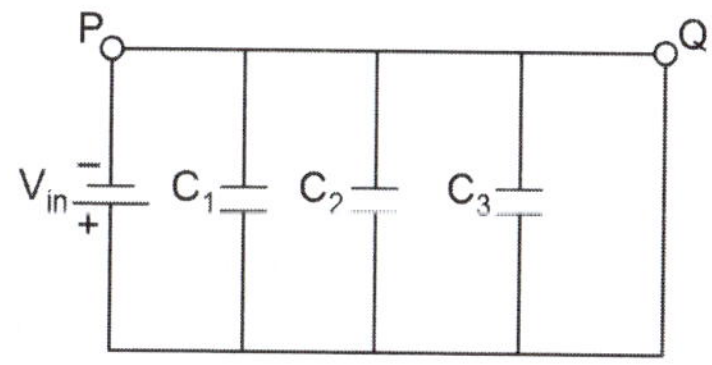

Fig. 4.34: Capacitor in parallel

Construction: Let us consider the following circuit in which three capacitors are connected in parallel with circuit. Placing the capacitors in parallel will increases the size of the capacitor's plates without increasing the distance between them. So the total capacitance of the parallel capacitor circuit is obtained by simply adding up the capacitance values of the individual capacitors.

The total capacitance is $C_T = C_1 + C_2 + C_3$

4.19.7 Current and Current Density

Current: In a simple electric circuit, the current is nothing but simple flow of electrons and it is represented by the letter *i*. The SI unit of current is *ampere,* which is measured by the flow of electric charge through a surface at the rate of one coulomb (6.24×10^{18}) of charge per second. It can be measured by using ammeter. It is represented by

$$I = \frac{dQ}{dt}$$

According to physicists, current is considered to move from relatively positive to negative points which is against the direction of electron flow, and this is also known as conventional current.

Types of current: The current can be divided into two types.

Direct current: Direct current travels towards the same direction at all points which means unidirectional flow of electric charge, although the instantaneous magnitude can differ.

The flashlight or another appliance running on batteries are supplied with direct current. One advantage of direct current is that it is relatively cheap to change the voltage of the current. Furthermore, the unavoidable loss of energy that occurs when current is carried over long distances. Electrochemical cell is also an example for DC generation.

A battery is a good example of a DC power supply. Direct current may flow in a conductor and semiconductors, insulators, or even through a vacuum as in electron or as ion beams. The electric current flows in a constant direction and a term formerly used for this type of current is galvanic current as shown in Fig. 4.35.

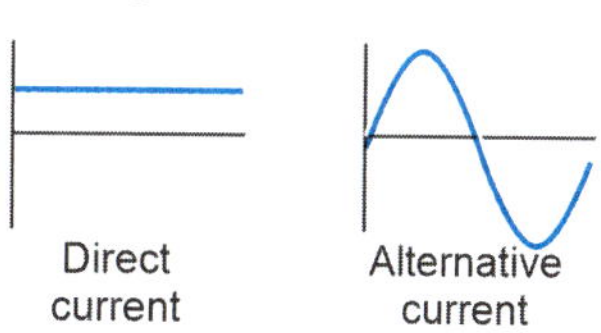

Fig. 4.35: Capacitor in parallel

By using of a rectifier circuit, we can get direct current from an alternating current supply, direct current may be converted into alternating current with the help of inverter or a motor-generator set.

Alternating current: The flow of charge carriers is towards the opposite direction periodically in an alternating current. The number of AC cycles per second is known as frequency and calculated in Hertz.

In physics, the electric current is the flow of electron and the direction of the flow of electrons shifts back and forth at regular intervals or cycles is called alternating current (AC) as shown in Fig. 4.36.

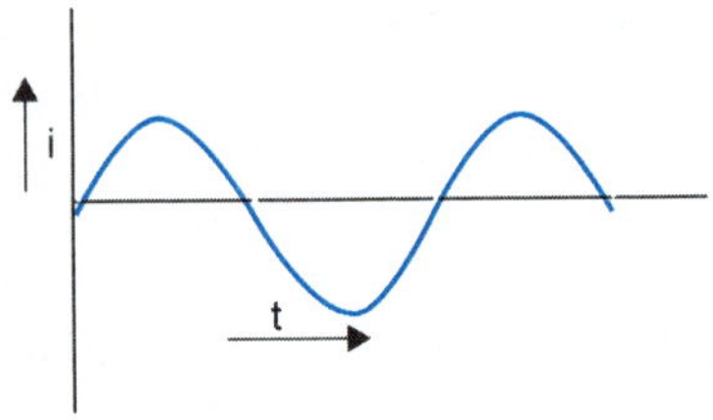

Fig. 4.36: Alternating current

50 cycles per second is the frequency of standard current used in India. This AC is used in India to supply power to houses, buildings, office, etc.

Waveforms: It has several forms, as long as the voltage and current are alternating. Sine wave is the usual waveform of alternating current as shown in Fig. 4.36. In certain applications, different waveforms are used, such as triangular or square waves. Examples of alternating current is audio and radio signals that are carried out by an electrical wires. These types of current carry information as AC signal of sound (audio) or images (video). These currents typically alternate at higher frequencies than those used in power transmission.

Table 4.1: Comparison table for AC and DC

	Alternating Current (AC)	Direct Current (DC)
Carrying amount of energy	It can be transfer to long distance between city for providing more power	It may loss energy when it carries for long distance
Direction of flow of electrons	Along the wire, rotating magnetism	Along the wire, steady magnetism
Frequency	Depending upon the country frequency may vary. The frequency of alternating current will be 50 Hz or 60 Hz in India	The frequency of direct current is zero
Direction	It reverses its direction while flowing in a circuit.	It flows only in one direction in the circuit
Current	It may varying with time in current magnitude	It has constant magnitude
Flow of electrons	Switches the directions—forward and backward	Steady in one direction or 'forward' only
Obtained from	AC generator	Battery or cell

Current density: In general, density is measure of mass per volume but here the current density is the amount of electric current that travel per unit cross-section area is called current density and expressed in amperes per square meter.

If the current in a conductor is more then the current density also raised. However, it may alters in different parts of an electrical conductor and it happens on when the alternating current conducts at higher frequencies. According to Faraday's law, the electric current always creates a magnetic field. If the current is more the magnetic field also increases. Varying AC or DC creates an electromagnetic field and this is the principle based on which signal propagation takes place.

It is a vector quantity which means it has both direction and magnitude. The electric current flowing through a solid having units of charge per unit time is calculated towards the direction perpendicular to the flow of direction.

Formula: The formula for current density is given as,

$$J = \frac{I}{A}$$

Where,

I = Current flowing through the conductor (in amp)

A = Area of cross-section (in m^2)

∴ Current density is expressed in A/m^2.

4.20 OHM'S LAW

Ohm's law deals with the relationship between the electric current and the potential difference in an ideal conductor. This relationship states that "the voltage across a conductor is directly proportional to the current flowing through it, provided all physical conditions and temperature remain constant".

The mathematically way of representation of current-voltage relationship can be written as,

$$V = IR$$

$$I = \frac{V}{R} \text{ or } R = \frac{V}{I}$$

The constant of proportionality is called the "resistance", *R*. where *V* is the potential difference between two points which include a resistance *R*. *I* is the current flowing through the resistance.

4.20.1 Resistance

When the electrons flow through the conductor, they are scattered by an atom in the conductor. The flow of electron is called current where the opposition of flow of current is called a resistance. The resistance of the circuit is also used to measure of electrons that flow in circuit. So that materials which are having a low resistance are called conductors and materials that have a very high resistance are called insulators. Some of the materials which have a moderate resistance and still allow some current to flow that type of material is called semiconductor.

Materials that we are using to make resisters like the ones type that can be used in experimental purposes. For example, if two or more resistors can be connected in series, connected one after another as shown in Fig. 4.37a or in parallel in Fig. 4.37b. If the current can split (i.e. there is more than one way for the current to flow and then rejoin is called resistor in parallel.

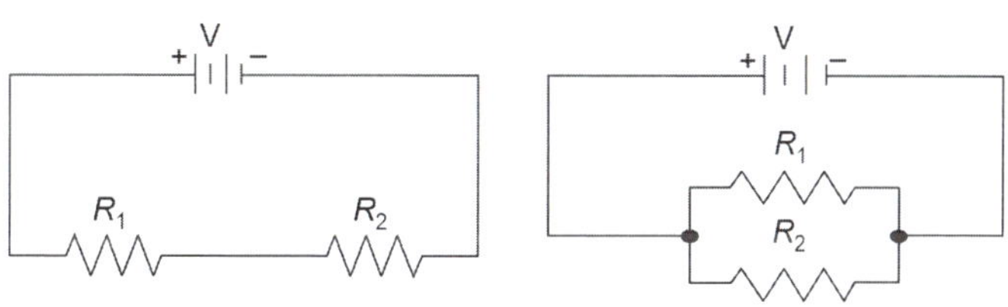

Fig. 4.37: Alternating current

Considered if two resistors R_1 and R_2 are connected in series in a circuit, therefore, the equivalent resistance R_S, is given by $R_S = R_1 + R_2$. Thus, the resistance is series in a circuit behaves like a single resistor with resistance R_S. In Fig. 4.37b, similarly those resistors are connected in parallel, the equivalent resistance in parallel circuit is given in Table 4.2.

Table 4.2: Series and parallel circuit equations

Series	Parallel
$V_s = V_1 + V_2$	$V_P = V_1 = V_2$
$I_s = I_1 = I_2$	$I_P = I_1 + I_2$
$R_s = R_1 + R_2$	$\frac{1}{R_P} = \frac{1}{R_1} + \frac{1}{R_2}$ or $R_P = \frac{R_1 R_2}{R_1 + R_2}$

Using these connection of resistors in series and parallel circuit, a complex circuit can be redesigned as a circuit with a single resistor.

4.21 KIRCHHOFF'S CIRCUIT LAW

Kirchhoff's circuit law plays a lead role in circuit analysis. The individual components of circuits are resistor, capacitor and inductor which are common tool for analyzing circuits.

Gustav Kirchhoff, a German physicist, who developed a pair or set of rules or laws in 1845. It deals with the conservation of current and energy in electrical circuits. These two rules are called Kirchhoff's circuit laws. In the first one of Kirchhoff's law deals with the current flowing around a closed circuit, *Kirchhoff's current law (KCL),* whereas the second law deals with the voltage sources present in a closed circuit, *Kirchhoff's Voltage Law (KVL).*

Kirchhoff's First Law: The Current Law (KCL)

According to this law, *total current or charge entering a junction or node is exactly equal to the charge leaving the node as it has no other place to go except to leave, as no charge is lost within the node.* Simply, the algebraic sum of all the currents entering a node and leaving it must be equal to zero.

$$I_{(exiting)} + I_{(entering)} = 0.$$

This idea by Kirchhoff is commonly known as the conservation of charge.

Let us consider the 3 currents entering the node, I_1, I_2 and I_3 are positive in value and the 2 currents leaving the node, I_4 and I_5 are negative in value as shown in Fig. 4.38. It can be depicted in equation,

$$I_1 + I_2 + I_3 - I_4 - I_5 = 0$$

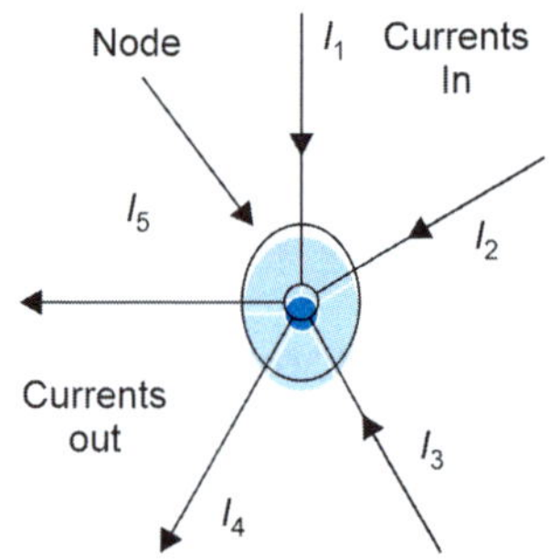

$I_1 + I_2 + I_3 - I_3 - I_5 = 0$

Fig. 4.38: Kirchhoff's current law

Node in an electrical circuit is called connection or junction of two or more current in a path. It may either in or out of a node in a closed circuit path. We can use Kirchhoff's current law for testing parallel circuits.

Kirchhoff's Second Law: The Voltage Law (KVL)

According to this law, *in any closed loop network, the total voltage around the loop is equal to the sum of all the voltage drops within the same loop* that is equal to zero. Simple, the algebraic sum of all voltages within the loop must be equal to zero. This is known as the conservation of energy defined by Kirchhoff.

Let us consider a starting point in the loop which continues in the same direction noting that all the directions of voltage drops, it may either positive or negative, and returning back to the same starting point as shown in Fig. 4.39. It is important to maintain the same

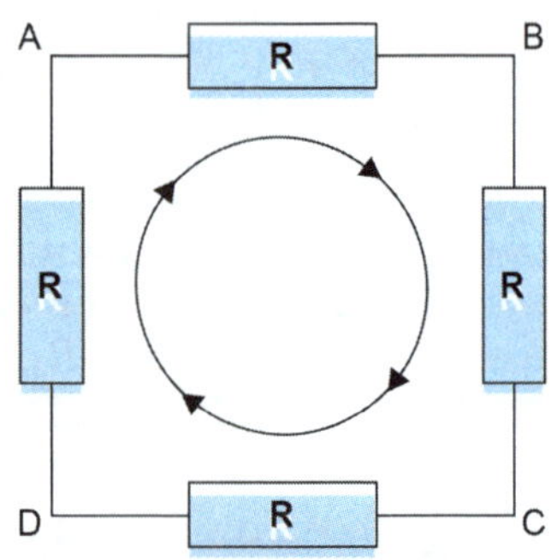

Fig. 4.39: Kirchhoff's voltage law

direction either clockwise or anti-clockwise or the final voltage sum will not be equal to zero. We can use Kirchhoff's voltage law for testing series circuits. Kirchhoff's circuit laws are used to test DC or AC circuits with terminologies and like node, paths, loops, etc.

4.22 EFFECTS OF ELECTRIC CURRENT

When current passes through the conductor that exhibits various effects. The main effects are heating, chemical and magnetic effects. These effects used in various purposes in day to day life.

i. **Heating effect:** As we known, when the current flow through the conductor that produces heat. But the principle behind the production of heat is unknown. In general, the flow of current (electron flow) is resisted in conductor that causes heat in it. Usually, the work has to be done to overcome the resistance which is converted into heat energy. According to Joule heating, there is conversion of electrical energy in to heating energy. This principle of heating effect is used in many home appliances like iron box, water heater, toaster, etc.

ii. **Chemical effect:** In physics, the flow of electrons is called current. But in case of electrolyte like copper sulphate solution, both the electron as well the positive copper ion also conducts current. The development of current through solutions is called '*electrolysis*' and the solution through which the electricity passes is called '*electrolyte*'. The anode which served as positive terminal inserted into the solution and the cathode that served as negative terminal. In the electrolysis experiment, copper wire is anode and carbon rod is cathode.

iii. **Magnetic effect of electricity:** According to Faraday's law of electromagnetic induction, when current flows to the conductor that develops a magnetic field which is perpendicular to the direction of current. It is called magnetic effect of current (*see* Section 4.10).

4.22.1 Heating Effect: Joule's Law

James Prescott Joule is an English Physicist who discovered the Joules' law and it is also called Jouel's effect, Joule-Lenz law or Joule's first law. This is the relationship between current, heat and resistance at a specific time, i.e. when a current flows through a material, it produces heat in it.

Joule's law: Joule's law states that when a current "i" amperes flows through a conductor of resistor "R" for "t" seconds that produce a heat "H" in it, then the amount of work done (conversion of electrical energy into heat energy) is equals to

$$\text{Work done} = \text{Heat } (H) = I^2Rt$$

Where, R and t are constant, the work done here is the amount of heat energy that converted from electricity which dissipates in the air.

Heating effect of current: Refer to Section 4.22 *about heat effect*, the current flows through a conductor that may be a wire, plate or cable produces heat and later it becomes hot. Because of applied current through a conductor, the applied electrical energy is converted into heat energy which increases the temperature on conductor.

As we know, electron flow is a current in the conductor, the drifting electrons in the substance collides with each other and the electrons of molecule atoms of the material. The collision of electrons produces heat. This effect is called the heating effect of the current.

Applications: Joule's law or heating effects of electric current are widely used in many domestic appliances and industrial purposes.

- Electric clothes iron
- Electric heaters, stoves, water heaters and heating elements
- Electric hot plate
- Electric welding
- Filament of incandescent lamps and light bulbs
- Food processing
- Fuses and fuse elements, etc.
- Hair dryers
- Soldering iron

Refer to Section 5.6. for seebeck effect, peltier effect and thompson effect.

4.23 MAGNETISM

Magnets are readily available material in domestics and workroom. These magnets are in use starting from common purposes like holding pictures on the refrigerator. Most of us are very familiar with properties of magnets and at the same time less familiar with the source of magnetism. So here come to play traditional concept of magnetism called magnetic field. It defines a volume of space where there is a change in energy with volume and it can be detected and measured. If the location of magnetic field which is detected are exiting or entering a material is known as magnetic pole.

For example, a bar magnet is considered as a dipole with a north and south pole. A field can be measured when it is leaving the dipole at north and returning to the south pole. In incident, if it breaks into two half then two magnets or dipoles are created. This sectioning and creation of dipoles can continue to the atomic level. Therefore, the source of magnetism lies in the basic building block of all matter in the atom.

The source of magnetism: As we know, all the matters in the universe is made up of atoms and sub-atomic particles that (*see* Section 2.2) are protons, neutrons and electrons. Protons and neutrons are present inside the nucleus and whereas the electrons are revolving around the nucleus. A magnetic field is formed when an electrical charge is in motion (*see* Section 4.9). The strength of this field is called the magnetic moment. So that all matter is made up of atoms, all materials are affected by an influence of magnetic field. Though, not all materials react in the same way, depends.

4.24 MAGNETIC INDUCTION (B)

Magnetic induction or electromagnetics is defined through its action on a moving charge and production of EMF. If a charge 'q' moving with velocity 'v' in a magnetic field that experiences a force 'F'. So that magnetic induction 'B' at a point is given by

$$F = qv \times B$$

The magnitude of,

$$B = \frac{F}{qv \sin \theta}$$

Here, 'θ' is the angle between v and B. Unit of B is weber/m^2. The magnetic field and its magnitude is defined by its induction properties.

4.25 MAGNETISATION (M)

When a material which is having magnetic property is placed in a magnetic field, the basic current loops in the materials arranged parallel to field direction. The material is now said to be magnetized and acquired a magnetic dipole moment.

Magnetisation M of the material is defined as induced dipole moment in a unit volume of magnetic material with unit of Am^{-1}. It is also called magnetic moment divided by the volume, then

$$M = \frac{m}{v} \quad ...(4.13)$$

Magnetisation M will be constant throughout the material which is magnetized and dipole in the same direction. In an unmagnetized material M will be Zero.

4.26 MAGNETIC SUSCEPTIBILITY (χ_m)

In a magnetic material, it found that the magnetisation 'M' is proportional to the magnetic field intensity 'H'.

$$\therefore \quad M \,\alpha\, H$$

$$\text{Or} \quad M = \chi_m H \quad ...(4.14)$$

Where, χ_m is called magnetic susceptibility of the material. It may also defined as the ratio between the magnetisation to its magnetic field density.

$$\therefore \quad \chi_m = \frac{M}{H} \quad ...(4.15)$$

Susceptibility 'χ_m' values can be determined by the type of magnetic material. If χ_m is positive, the material is called paramagnetic. If χ_m is negative, the material is said to be diamagnetic. In case of ferromagnetic material, the susceptibility χ_m value is positive and very large. Though in ferromagnetic materials, M is not exactly proportional to H and χ_m is not a constant.

4.27 MAGNETIC PERMEABILITY (μ)

Let us consider the an isotropic medium,

$$B = \mu_0 (H + M)$$
$$= \mu_0 (H + \chi_m H)$$
$$= \mu_0 (1 + \chi_m)H$$
$$\therefore \quad B = \mu H \quad ...(4.16)$$

From Eq. (4.16), μ is the magnetic permeability of the material. It is defined as the ratio between magnetic induction to the magnetisation in intensity field.

$$\therefore \quad \mu = \frac{B}{H} \quad ...(4.17)$$

For vacuum $\chi_m = 0$ and $\mu = \mu_0$.

Hence magnetic induction in vacuum is $B_0 = \mu_0$ H.

The ratio: $\frac{B}{B_0} = \frac{\propto}{\propto_0}$ is called the relative permeability μ_r obviously $\mu_r = 1 + \chi_m$.

Depending upon the relative permeability μ_r, the material can be classified as

Diamagnetism: $\mu_r < 1$ (for Bi, $\mu_r = 1 - 0.00017$)

Paramagnetism: $\mu_r < 1$ (for Al, $\mu_r = 1 - 0.00002$)

Ferromagnetism: $\mu_r < 1$ (for pure Fe, $\mu_r = 1 - 200{,}000$)

4.28 MAGNETIC MATERIALS

All matters of universe exhibit magnetic behavioural property. Matters like this made up of charged particles and according to atom model, arrangement of particles that determines the magnetic properties of the material. Depends, this may exhibit the effect of magnetic dipole due to applied magnetic field. Based on the magnetic properties, materials can be classified as follows.

Usually, the material is placed in an uniform magnetic field and behaviour of the magnetic field is varied. There are three major kinds of magnetic behaviour:

i. Diamagnetic materials
ii. Paramagnetic materials
iii. Ferromagnetic materials

4.28.1 Diamagnetism

Material of this kind is hardly magnetized when it is placed in a magnetic field. Its one property is of dipole, that tends to align in opposite direction to the applied magnetic field. So that it can repel the external field around it. In effect, they produce an internal magnetic field that opposes the applied field.

This opposing field disappears as soon as the external field is removed.

Properties of diamagnetic materials

i. Relative permeability μ_r is less than 1.
ii. Susceptibility 'χ_m' of this material has less negative value and it is independent of temperature.
iii. When it is placed in an external magnetic field, it gets magnetized in opposite direction and has capability to move away from external magnetic field.
iv. When it is suspended freely, they set in perpendicular to the field.
v. *Bi, Zn, Ag, Pb Au*, etc. are belongs to diamagnetic materials.

4.28.2 Para-magnetism

Material of this kind is weakly magnetized when it is placed in a magnetic field, magnetic dipoles of the tend to align along the applied magnetic field. So that it can reinforcing the applied magnetic field and this material is attracted by a magnet. The magnetization of the material disappears as soon as the external field is removed.

This type of material was first discovered by Madam Curie and it is dependent on the external magnetic field (*B*) and temperature *T* as:

$$\vec{M} = C\frac{\vec{B}}{T}$$

Where, C = Curie constant

Properties of paramagnetic materials

i. Relative permeability μ_r is greater than 1.
ii. Susceptibility 'χ_m' of this material has positive value.
iii. It is temperature dependent material and χ_m will decrease with increase in temperature.
iv. When it is placed in an external magnetic field, it gets magnetized in same direction and has capability to enhance the magnetic field of applied external magnetic field.
v. When it is suspended freely, they set in parallel to the field.

vi. Liquid oxygen, sodium, platinum, salts of iron, chromium, manganese, nickel, etc. are the examples of this materials.

4.28.3 Ferromagnetic

Materials of this kind exhibit the strongest magnetic behaviour. Magnetic dipoles in these materials are arranged inside the material where the arrangements of individual magnetic dipoles plays a role. If they are perfect in same direction then it can produce strong magnetic fields. Normally, material of this kind is usually randomly arranged, and thus, the magnetic fields are cancelled by another dipole and the entire material does not show any magnetic behaviour.

But when a material is placed in an external field, it will be reoriented themselves to reinforce the external field and produce a strong internal magnetic field. When the field is removed, it will try to be in same orientation towards in the direction of the magnetic field. So those magnetic properties are persisting even when the external field disappears. This property is used to produce permanent magnets that we use everyday.

Properties of ferromagnetic materials

i. Permeability μ_r and susceptibility 'χ_m' of these materials are very large.

ii. When it is placed in an external magnetic field, it gets magnetized strongly towards in the direction of magnetic field and they are strongly attracted by magnets.

iii. It is temperature depended material, when the temperature of the material increases, the value of susceptibility 'χ_m' decreases and above certain value of temperature is also called curie temperature, ferromagnetic material become paramagnetic materials.

iv. This material also exhibits the phenomenon of hysteresis.

v. Iron, nickel and cobalt are some examples of ferromagnetic materials.

vi. There exist two other types of magnetism which are very closely related to ferromagnetism. These are antiferromagnetism and ferrimagnetism.

4.28.4 Antiferromagnetism

Ferromagnetic materials have special type of magnetism which exhibit that in magnetized state. During this state, all the dipoles are aligned in the same direction (Fig. 4.40a). But there are some substances which show different types of magnetism. In antiferromagnetic materials like C_r and M_nO, the dipoles have equal moments, but the moments of neighboring dipoles in opposite direction (Fig. 4.40b). Thus, the moments balance each other, resulting in zero magnetisation.

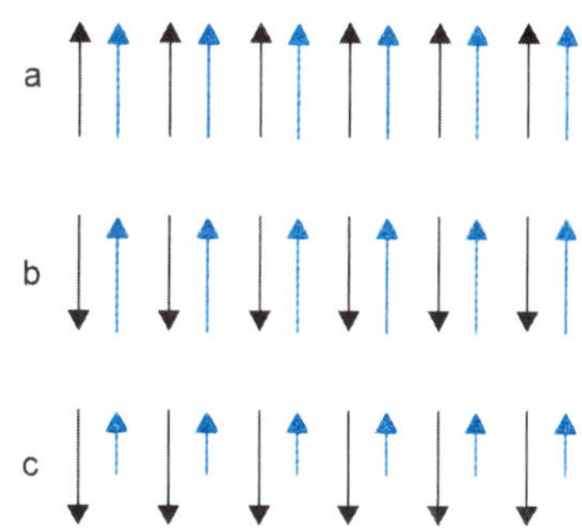

Fig. 4.40a to c: Types of magnetism

Antiferromagnetic material is depending upon the temperature, it exhibits a special behaviour in an applied magnetic field. At very low temperatures, it does not response to external field, because material dipoles are maintained rigidly. At higher temperature, some atoms break free of the orderly arrangement and align with the external field. The temperature at which the alignment of magnetism become weak in the solid is called Neel temperature. Above this temperature, the material become paramagnetic.

4.28.5 Ferrimagnetism

Ferrimagnetic material is a type of permanent magnetic material that presents in solids where the magnetic field is associated with individual atoms. Ferrite material is belonging to ferrimagnetism. It is also observed that the magnetic moments in the material are aligned in parallel and anti-parallel directions in uneven numbers.

Ferrimagnetism is having similar behaviour like ferromagnetism. It exhibits all the ferromagnetic behavior, magnetization, Curie temperatures and hysteresis. But this will differ in between the magnetic property by its different magnetic fields ordering (Fig. 4.40c).

4.29 MAGNETIC EFFECT OF CURRENT

A magnetic field is a type of force that created by electric charges and magnetic dipoles. Like the other effects of magnetic field, the electric current also produces some magnetic effect. This magnetic effect of electric current is known as electromagnetic effect.

Let us consider a compass box, when it is placed near to current carrying conductor the needle of compass gets deflected magnetic field which is created by current. This shows electric current produces a magnetic effect.

4.30 MAGNETIC FIELD AROUND A STRAIGHT CONDUCTOR CARRYING CURRENT

When a current passes through the conductor that creates magnetic field in the form of concentric circles around it. Magnetic field can be shown by magnetic field lines (Fig. 4.41a). The direction of magnetic field in a conductor will depend upon the direction of current flow. If the direction of current flow is reversed and the direction of magnetic field also reversed.

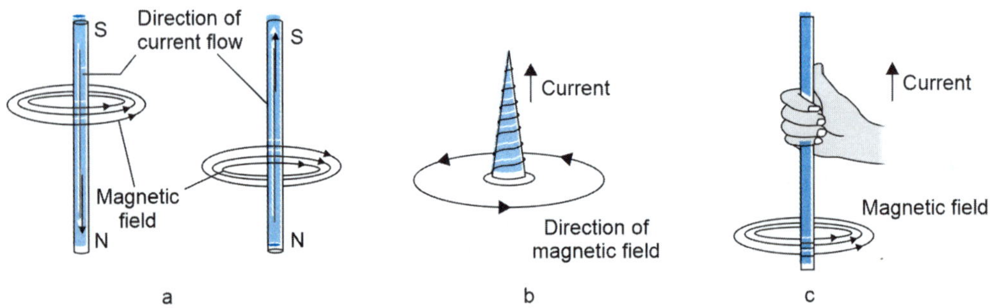

Fig. 4.41: a. Magnetic field in a conductor, b. Maxwell's corkscrew rule, c. Right hand clasp rule

Let us consider a current carrying conductor with current flow from one direction, for example, south to north, and it is observed that the direction of magnetic field will be anticlockwise. If the current flow in opposite direction like from north to south, the direction of magnetic field will be clockwise.

Maxwell's corkscrew rule: If a right-handed screw is turned to forward along the direction of conductor, the direction of rotation of the screw shows the direction of the lines of force (Fig. 4.41b).

Right hand clasp rule: Hold the conductor in the right hand with your thumb projecting up which is the direction of the current in the conductor. Then the direction of the rest of the finger gives the direction of the magnetic produced by the conductor (Fig. 4.41c).

4.31 BIOT SAVART LAW

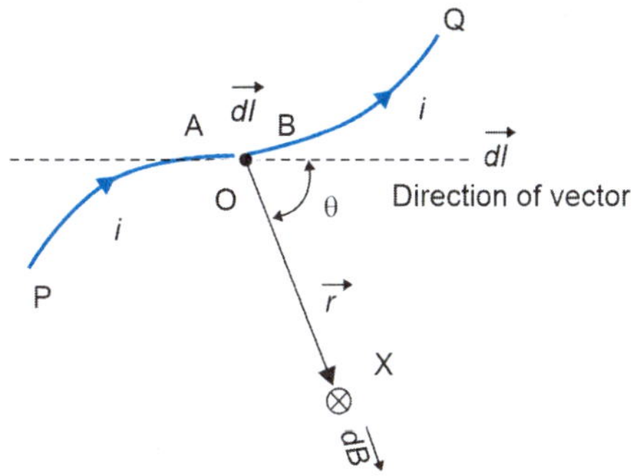

Fig. 4.42: Biot-Savart law

Let us consider a conductor PQ carrying a current '*i*' (Fig. 4.42). Consider an element AB of length *dl*, O is the midpoint of AB. *X* is a point at a distance *r* from O. θ is the angle between *dl* and *r*. Magnetic induction *dB* at point *X* due to the current element *dl* is

$dB = \left(\frac{\mu_0}{4\pi}\right)\frac{i(dl \times \hat{r})}{r^2}$ where $\hat{r}$ is a unit vector along *r*

The magnitude of *dB* is
The magnitude of *dB* is

$$dB = \left(\frac{\mu_0}{4\pi}\right)\frac{i\,dl\sin\theta}{r^2} \qquad ...(4.18)$$

The direction of dB is the vector of $dl \times r$.

The total magnetic induction *B* at P due to the current flowing in the entire length of the conductor is given by

$$B = \int dB = \left(\frac{\mu_0}{4\pi}\right)\int i = \left(\frac{dl \times \hat{r}}{r^2}\right) \qquad ...(4.19)$$

In vaccum, *B* is related to *H* (magnetic field intensity) is given by

$$B = \mu_0 H \qquad ...(4.20)$$

Where, μ_0 is a constant, called permeability of free space.

4.32 AMPERE'S CIRCUITAL LAW

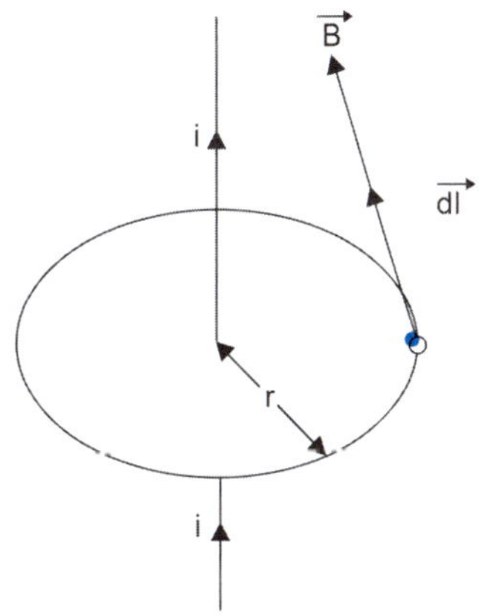

Fig. 4.43: Ampere's circuit law

The line integral $\oint B \cdot dl$ for a closed curve is the product of permeability of free space to the net current '*i*' flow through the area bounded by the curve is given below

$$\oint B \cdot dl = \mu_0 i, \qquad ...(4.21)$$

This is called Ampere's circuit law.

Let us consider a long straight conductor carries a current '*i*' which is perpendicular to the magnetic field *B* as shown in Fig. 4.43. According to Biot Savart law, the magnitude of the magnetic induction *B* at a distance *r* is given by,

$$B = \frac{\mu_0 i}{2\pi r} \qquad ...(4.22)$$

All point in the circle of magnetic field, B has a constant magnitude *B* and *dl* which is always tangential to the path of integration, therefore,

$$\oint B \cdot dl = \oint B dl = B\oint dl = (B)(2\pi r)$$

Here, $2\pi r = \oint dl$ is the circumference of the circle. Substitute the value of B from Eq. (4.22) we get,

$$\oint B \cdot dl = \frac{\mu_0 i}{2\pi r}(2\pi r) = \mu_0 i \qquad ...(4.23)$$

Therefore, $\oint B \cdot dl$ is the integral of magnetic field density along an imaginary closed path is equal to the product of current enclosed by the path and permeability of the medium is called Ampere's law.

4.33 SOLENOID

A solenoid is coil of wire which is used in the housing and a moveable needle to generate a strong magnetic field inside the coil is called solenoid. When an electrical current is passed through it, a magnetic field is generated around the coil that produce a force to pull the needle. If N is the number of turns of the solenoid coil wire, then a greater number of coil produces strong magnetic field. So simply, a solenoid can pull the needle easily and can converts electrical energy into mechanical work. The main purpose of solenoid is to impede changes in electric current.

The magnetic field of a solenoid is given by,

$$B = \frac{\mu_0 i_n}{l} \quad \ldots(4.24)$$

Where,

n = number of turns in the solenoid coil of wire

i = current flow through the coil

l = length of the solenoid coil.

From Eq. (4.24), the magnetic field is produced in the coil is directly proportional to the current flow and the number of turns per length of the coil.

4.33.1 Right-Hand Palm Rule

When a current flow through the solenoid conductor, the direction of the magnetic field due to circular closed loops (solenoid) is given by right-hand palm rule. This the method in which we can find the product of two vectors and also used to find the force on a current carrying conductor in a magnetic field.

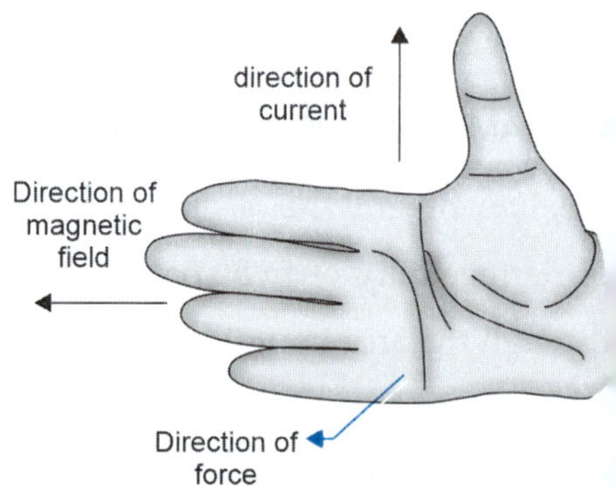

Fig. 4.44: Right-hand palm rule

Stretch the fingers of right hand which should be right angle to thumb finger as shown in Fig. 4.44. Then if the fingers of right hand is pointing out the direction of magnetic field and the thumb finger is pointing the direction of current flow I, then the palm of right hand pointing out the direction of force. This is called right-hand palm rule.

4.33.2 End Rule

When a current flow through a conductor of solenoid which is suspended freely, at first it oscillates and comes to rest like a bar magnet positioning along north–south direction. The magnetic polarity of the current carrying solenoid is given by end rule.

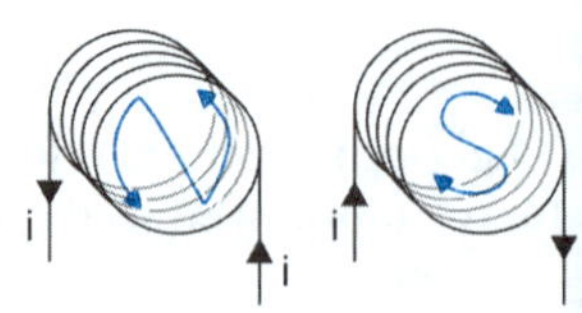

Fig. 4.45: End rule

It states that if current flows in the solenoid conductor in clockwise direction the nearer end of solenoid depicts south pole and the other end is north pole (Fig. 4.45). If the current flows in the anticlockwise direction the nearer end of solenoid depicts north pole and the farthest end is south pole

4.34 GALVANOMETER-MOVING COIL GALVANOMETER

In physics, the galvanometer is an electromechanical device that used to detect the presence of small current or voltage and its magnitude. In practical purposes, it can be kept in the bridges and potentiometer where it can indicate the null deflection or zero current. Usually, the strength of the current is indicated by a magnetic needle's movement

Principle: When a current is passed through a coil in the presence of magnetic field, the coil experiences a torque which is proportional to the current.

Construction: The construction of the galvanometer is given in Fig. 4.46. The suspension, moving coil and stable magnet are the main parts of the galvanometer.

Suspension: The coil is balanced with a help of plain ribbon. The ribbon is made up of conductor that supplies the current to the coil. The other coil carries the current in the lower suspension and torque effect can be negligible. The upper suspension is made up of with a gold wire or copper wire in the ribbon form. But ductility of wire is not extremely tough. So, the galvanometer handles carefully without any stress.

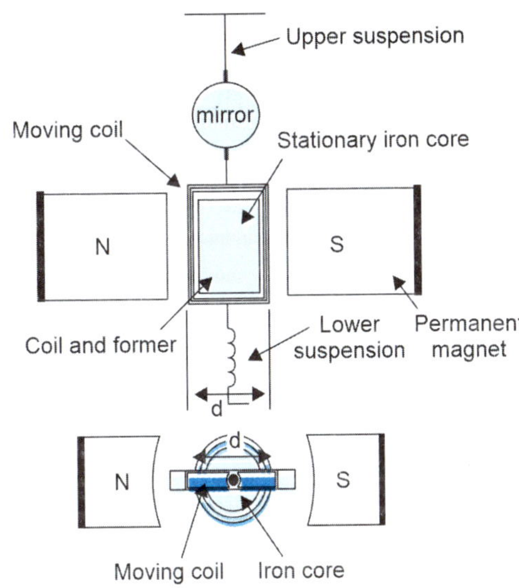

Fig. 4.46: Suspension

Moving coil: Current-carrying element of the galvanometer is the moving coil. This coil is in rectangular shape with the number of twists of copper wire. This coil made to move freely between the poles of stable magnet. The iron core magnetic flux flow is relatively low which also called low reluctance flux and, therefore, gives the tough magnetic field for the twist to move in.

Mirror: The one of the main part of suspension in galvanometer is a little mirror that works on passing the ray of light which is located on the scale and the deflection can be measured.

Torsion head: The location of the coil is mainly control by the torsion head and can be used to set zero marking in the scale.

Working: When the current flow in the circuit where a galvanometer is connected, then the flow of current will be facilitate in the coil. A magnetic field is influenced on the coil, then a deflecting torque works upon it. Due to this torque, a coil of galvanometer starts revolving from its place.

Once the coil starts spin, that delivers a twist and tries to restore torque, after that it resists the revolving of coil. The coil's rotation angle is directly proportional to the torque. When the current passes to the circuit, produces a torque that tries to normal restore position. This is equivalent to the deflecting torque. A galvanometer is mainly used in various electrical circuits for detecting current as well as in experiments to decide this null point.

If the galvanometer is overloaded with current, then the pointer might strike and pin will not move because of very large deflection. So, the coil in the galvanometer melts and may blaze because of extreme heat. This can be avoided by using a wide wire otherwise by connecting it to a copper strip in parallel with its coil which is known as a shunt. When compared to the coil's resistance, its resistance is extremely small. Thus, most of the current flow supplies through the shunt and some amount of current supplied through the coil.

Advantages

i. It will not affect by a strong magnetic field.
ii. Precise and reliable.
iii. Scales are uniform.

Disadvantages

i. Change within restoring torque by the change in temperature.
ii. Limitations in AC quantities measurement.
iii. Overload can spoil the galvanometer.
iv. Restoring torque is hard.

Applications

i. It can determine the voltage between two points.
ii. Null point and current can be determined in the circuit
iii. Used in a film camera to get the readings of photoresistor.
iv. Used in control systems, laser engraving, laser TVs, laser sintering, laser displays, etc.
iv. Used in the CD/DVD players for controlling the position of head.
v. Used to detect the flow of current's direction in a circuit.

4.35 CONVERSION OF GALVANOMETER INTO AN AMMETER

A galvanometer is an electromechanical device which is used to detect the flow of current in a circuit. The deflection in the meter is directly proportional to the current, the galvanometer scale deflection is not shown in ampere. It is a very sensitive instrument, if a large load of current passed on it that causes damage to the coil.

A galvanometer can be converted into an ammeter by connecting a low resistance to the cicuit with galvanometer in parallel. By doing this, when large current flows in a circuit small portion of current that passes through the galvanometer and the remaining larger portion of the current passes through the low resistance. The low resistance connected in parallel with the galvanometer is called shunt resistance. The scale is marked in ampere. So that it measure the current flow in the circuit only when a galvanometer is connected in series as shown in Fig. 4.47.

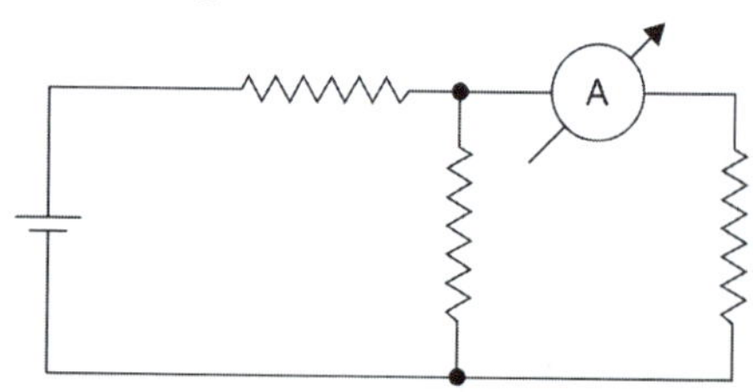

Fig. 4.47: Circuit diagram for conversion of galvanometer into ammeter

4.36 CONVERSION OF GALVANOMETER INTO A VOLTMETER

A voltmeter is an electrical device that can be used for measuring potential difference (the voltage) between two wires, usually across a circuit or a group of elements. A voltmeter is always connected in parallel with the element(s). This can be achieved by connecting a galvanometer with a high resistance in series with it. The scale is calibrated in volt. The value of the resistance connected in series decides the range of the volt in galvanometer as shown in Fig. 4.48.

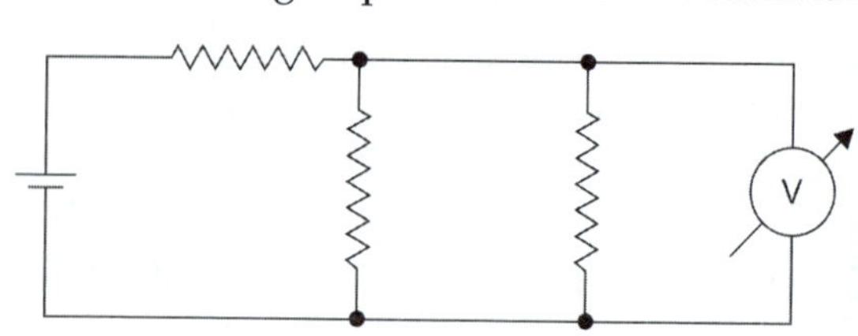

Fig. 4.48: Circuit diagram for conversion of galvanometer into voltmeter

4.37 MULTIMETER

The multimeter is a device used to measure voltage, current and resistance in a electrical and electronical instruments. In case of electrical instruments which consists of wire and fuses, it can be used to check the continuity of wire, fuses and other parts of devices. It can be used to check and testing transistors and capacitors in electronic devices.

It is a common analytical tool for experts and professionals in the electrical and electronic industries. In early days, normal needle based analog meters were in use and it has some disadvantages that it can be used only single task meters like voltmeter in volts or ammeter in amps or ohm meter in ohms. Later it is replaced by a digital multimeter (DMM) in 1977. It is used to measure two or more electrical values simultaneously.

The digital multimeter displays four components, they are:

i. **Display:** Showing the measured values of multimeter.

ii. **Buttons:** Ideal selection of various functions depends model will vary.

iii. **Dial *(like regulator or rotary switch)*:** For selecting primary measurement values (volts, amps, ohms).

iv. **Input jacks:** Test leads can be connected.

The circuit can be represented based on Ohm's law ($V=IR$) by a schematic diagram as shown in Fig. 4.49.

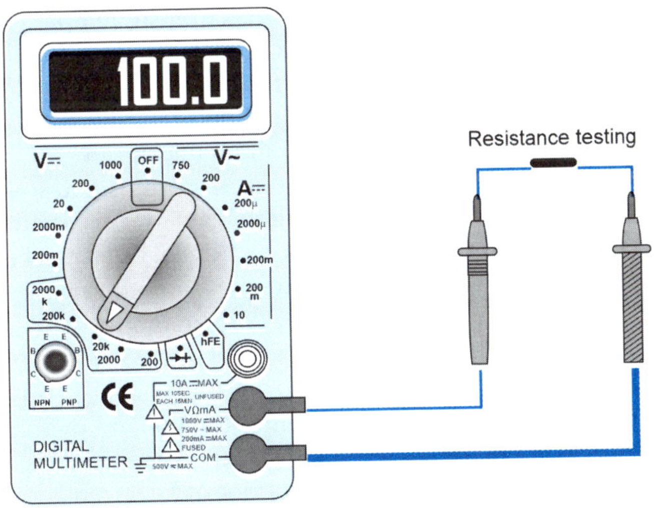

Fig. 4.49: Multimeter

Chapter

5

Heat and Thermodynamics

5.1 INTRODUCTION TO THERMODYNAMICS

Thermodynamics is one of the main branches in physics and engineering. It is also dealing on the study and application of the thermal energy which is also called internal energy of systems. One of the major concepts of thermodynamics is temperature and its measurements. To understand better, we need know thermal properties of matters and the processes.

Generally, in physics we need to know the notions of heat, temperature, work, etc. for betterment of knowledge on thermodynamics. At first, it was considered as a fine invisible fluid that filling up the pores of a material. When a hot body and a cold body are placed together, the form of energy is transferred from hot body to cold body till it become hot. This form of energy is called heat and it is also valued in calories.

Heat: Heat is the form of energy transferred between systems and it is mainly due to a temperature difference between them. According the law of conservation of energy, heat is conserved which means it cannot be created or destroyed. But it can be transferred from one place to another and can also be converted to one from other forms of energy.

For example, a light bulb can be used in that conversion takes place from electrical energy to electromagnetic radiation energy, called light and when this energy is absorbed by any of surface with matter that convert it into a heat. So, energy cannot be destroyed.

Temperature: When the heat energy is become measurable and quantified by changing it average kinetic energy of molecules of matters, then it is called temperature and it is a measurable physical property of matters. The amount of heat transferred by a substance depends on the speed and number of atoms or molecules in motion. The faster the atoms or molecules move, the higher the temperature, and the more atoms or molecules that are in motion, the greater the quantity of heat they transfer. It is measured in Celsius (°C), kelvin (K), Fahrenheit (°F), or Rankine (R)

5.2 THERMAL EQUILIBRIUM

A very important concept of thermodynamics is thermodynamic equilibrium. In general, the term equilibrium means in mechanics that the net external force and torque on a system is zero. But the term 'equilibrium' in thermodynamics mentions about the state of a system. The state of equilibrium of system that do not change in time. For example, a gas inside a closed chamber which is completely insulated from the surroundings with constant pressure, volume, temperature, and mass. This state does not change with time, is in a state of thermodynamic equilibrium.

5.3 LAW OF THERMODYNAMICS

There are four laws of thermodynamics which define the fundamental physical quantities such as temperature, energy, and entropy of a system at thermal equilibrium. These laws define the quantities and how it behaves under various circumstances.

The four laws of thermodynamics are:

i. Zeroth law of thermodynamics
ii. The first law of thermodynamics
iii. The second law of thermodynamics
iv. The third law of thermodynamics

5.3.1 Zeroth Law of Thermodynamics

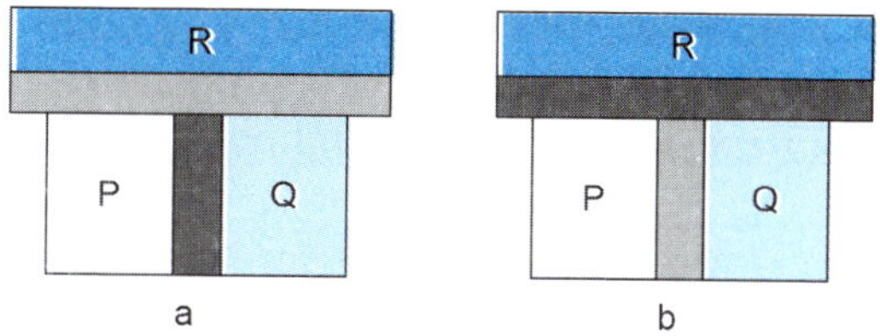

Fig. 5.1: Zeroth law of thermodynamics

Let us consider the two systems P and Q that are separated by an adiabatic wall (*heat does not enter or leave the wall of system, mentioned blue colour*), when these systems are in contact with a third system or equilibrium with *R* through a conducting wall (*green colour*) as shown in Fig. 5.1a. The states of macroscopic variable of P and Q will be altered when in contact with R. If the adiabatic wall is replaced with conducting wall and R is insulated from P and Q with adiabatic wall as shown in Fig. 5.1b. At this state, P and Q found to be in thermal equilibrium with each other. This process is called as basis of the zeroth law of thermodynamics, which states that *two systems in thermal equilibrium with a third system separately are in thermal equilibrium with each other*. This law helps define the notion of temperature.

5.3.1.1 Heat and Internal Energy

The zeroth law of thermodynamics is directed us to understand better about the concept of temperature. Temperature is a tool to determine the hotness of a body and the flow of heat when two bodies are in thermal contact. Usually, heat can transfer from at high temperature region to the lower temperature. When these two bodies are at equilibrium, the flow of heat stops.

The notion of internal energy of system can be explained with a large number of molecules in a bulk system. The internal energy which is the sum of the kinetic energy and potential energy of these molecules are sharing energy with a condition of centre of mass of the system is at rest in the frame of reference. But the internal energy of the system is depending on the state of the system and it does not depend on how the state of system formed.

5.3.2 First Law of Thermodynamics

The first law of thermodynamics, also known as law of conservation of energy which states that energy can neither be created nor to be destroyed; energy can only be transferred or changed from one form to another. For example, if you switch on a light, it would produce energy where its electrical energy is converted.

The first law of thermodynamics can be expressed in the form of energy that any change in the internal energy (°E) of a system is given by the sum of the heat (Q) that flows across its boundaries and the work (W) is done on the system by its surroundings:

therefore, $$\Delta E = Q + W \qquad ...(5.1)$$

from the above equation, there are two kind of process, heat and work, which can lead to a change in the internal energy of the system. Since both heat and work can be measured quantitatively. According to the interrelation between heat and work, any change in the energy of a system must result in a corresponding change in the energy of the surroundings that outside of system. In other words, energy cannot be created or destroyed. If heat flows into a system or the surroundings do work on it, the internal energy increases and the sign of Q and W are positive. On the other words, heat flow out of the system or work is done by the system (on the surroundings) and it will be expressed as the internal energy of the system so that Q and W will be negative.

5.3.2.1 Specific Heat Capacity

If a small amount of heat ΔQ is given to a substance that produces a changes in small amount of temperature ΔT of substance. Then the final temperature of substance is $T + \Delta T$. It can be expressed as follows.

$$S = \frac{\Delta Q}{\Delta T} \qquad ...(5.2)$$

The amount of heat given to a substance will be vary with its mass. So, the heat capacity S is also proportional to the mass of the substance and in addition it is depending on temperature, i.e. a different quantity of heat may require to increase the temperature of substance. To define a constant of specific heat of substance and independent of its amount, S could be expressed as,

$$S = \frac{S}{m} = \left(\frac{1}{m}\right)\frac{\Delta Q}{\Delta T} \qquad ...(5.3)$$

From the above equation, S is known as the specific heat capacity of the substance. It may vary depend on the nature of the substance and its temperature. The unit of specific heat capacity is $J\ \mathrm{kg}^{-1}\ \mathrm{K}^{-1}$.

If the amount of substance is expressed in terms of moles μ, so heat capacity per mole of the substance is given by

$$C = \frac{S}{\mu} = \frac{1}{\mu}\frac{\Delta Q}{\Delta T} \qquad ...(5.4)$$

C is known as molar specific heat capacity of the substance. If C is independent of the amount of substance, but it depends on the nature of the substance, its temperature and the conditions in which heat is supplied. So that the unit of C can be expressed as $J\ \mathrm{mol}^{-1}\ K^{-1}$.

From the Eq. (5.3) the specific heat capacity of substance, defined as "the amount of heat is required to raise the temperature of 1 gram of substance by 1°C."

Thus, specific heat values of a substance can be determined by a small practice, take two materials of different temperatures which are placed together in contact. As we know, the heat flows from the higher temperature region to the lower temperature region of material. According to the law of conservation of energy, the heat lost in the material is equal to the same heat that gained by the another material. Based on that, heat gained by the material that has increase in temperature which is equal to the temperature that lost by another substance in contact. It is also observed that heat gained by the substance may vary in difference in temperature and substance. So different substances have different specific heat and heat capacity S of a substance is the quantity of the heat required to raise the temperature of the whole substance by one degree. If the mass of the substance is unity then the heat capacity is called specific heat capacity or the specific heat.

From Eq. (5.3) and (5.4), specific heat capacity formula is given by

$$Q = Cm\,\Delta T \quad ...(5.5)$$

Where

Q = quantity of heat absorbed by a body

m = mass of substance

T = rise in temperature

C = specific heat capacity of a substance depends on the nature of the material of substance. SI unit of specific heat is J kg^{-1} K^{-1} *(mass is kg not in mole).*

5.3.2.2 Thermodynamic State Variables and Equation of State

The word equilibrium in thermodynamics is described by precise values of some macroscopical variables, called state variables. For example, in gas thermodynamically, an equilibrium state of a gas is stated by the different values of pressure, volume, temperature, mass and also its composition.

At the same time, a thermodynamical system is not always in equilibrium. In case, if a gas allowed it to expand freely which is not an equilibrium state. This is due to rapid expansion of gas that produce a state of non-equilibrium. Similarly, if the pressure and temperature are not uniform in the chemical reaction of a mixture of gases which undergoes an explosion. Later, the mixture of gas reaches the state of uniform temperature and pressure and comes to thermal and mechanical equilibrium with its surroundings as shown in Fig. 5.2.

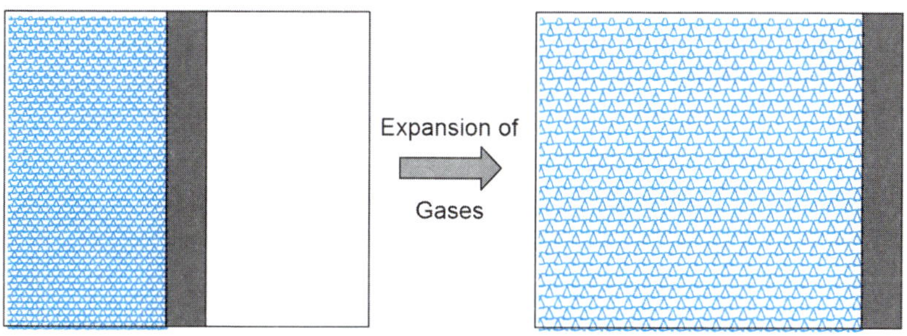

Fig. 5.2: State of thermodynamic system

5.3.2.3 Thermodynamic Processes

In thermodynamics, a process in which heat energy is transferred within a system or in between systems is called thermodynamic processes.

There are four types of thermodynamic processes. These are also classified into group process called pairs in which each variable held constant and one is a conjugate pair.

Pressure–volume:

(1) Isobaric and (2) Isochoric

Temperature–entropy:

(1) Isothermal and (2) Adiabatic

1. **Isobaric process:** A thermodynamic isobaric process occurs when pressure is maintained at constant. For example, a movable piston in a cylinder that the system is dynamically connected, by a movable boundary, to a constant-pressure reservoir.
2. **Isochoric process:** A thermodynamic isochoric process occurs when volume is maintained at constant. The resultant mechanical PV work done by the system will be zero. So it is also called an isometric process or constant-volume process. The term isochoric means "iso"—"constant" and "choric"—"space" or "volume." Where pressure and volume maintained constant.

   For example, in a closed tin box placed into a fire. In that instant, the tin box will not expand, and the only change is gaining in internal energy, this causes increase in temperature and pressure inside the tin where the volume is maintained at constant. Mathematically it is defined as,

   $$\delta Q = dU \qquad \text{...(5.6)}$$

   From the above equ (5.1), the total heat given or rejected is also equal to the increase or decrease in the internal energy of the system.
3. **Isothermal process:** A thermodynamic isothermal process occurs only when the temperature is kept constant as the pressure increases during compression. At the same time, it is not possible to build a device that will compress isothermally, isothermal performance is attained only when the number of intercoolers or other cooling devices is increased. In practice, isothermal compression cannot be attained because if a system is in contact with a thermal reservoir from outside, then, to maintain thermal equilibrium, the system slowly adjusts itself with the temperature of the reservoir through heat exchange. That means during an isothermal process there is a change in internal energy, heat energy, and work, even though the temperature remains the same.

   For an ideal gas, $$PV = nRT \qquad \text{...(5.7)}$$

   where, n = no. of moles (constant),

   R = universal gas constant,

   T = constant for isothermal process.

   Therefore, from Eq. (1), PV = constant.
4. **Adiabatic process:** A thermodynamic adiabatic process in which there is no exchange of heat from the system to its surrounding neither expansion nor compression. It means the process can be either reversible or irreversible. The essential conditions for the adiabatic process are as follows:
   a. The system should be insulated from the surrounding.
   b. The process must be carried out quickly so that there is enough time for heat transfer to take place.

5.3.2.4 Heat Engines

It is a type of device that helps to convert heat to work in which system is made to undergo a cyclic process. For example, in a system with a mixture of fuel vapor and air in an engine are the working substances of that system. There are several processes on working substance which undergoes through a cycle. In some of these processes, it absorbs a total amount of heat Q_1 from an external reservoir at some high temperature T_1. In some other processes of the cycle, the working substance releases a total amount of heat Q_2 to an external reservoir at some lower temperature T_2.

The system work done (W) in a cycle is transferred to the environment through a moving piston in a cylinder that transfers mechanical energy to the wheels of a vehicle via a shaft. The features of a heat engine are shown in Fig. 5.3.

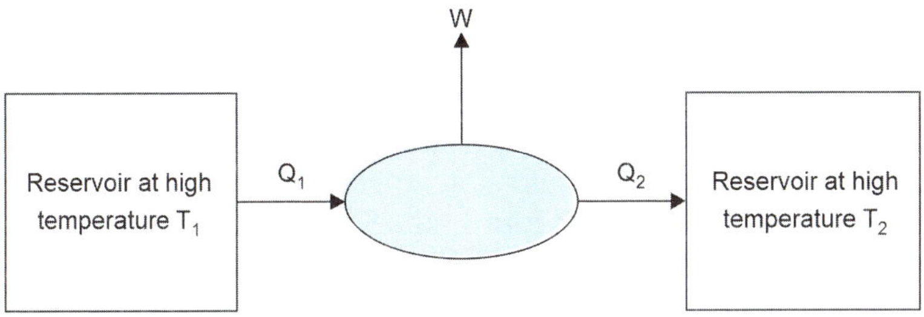

Fig. 5.3: Schematic diagram of heat engine

The efficiency (η) of a heat engine is defined by

$$\eta = \left(\frac{W}{Q_1}\right) \qquad \text{...(5.8)}$$

where, Q_1 is the heat input, i.e. the heat absorbed by the system in one complete cycle and W is the work done on the environment in a cycle. In a cycle, a certain amount of heat (Q_2) may be rejected to the environment. Then, according to the first law of thermodynamics, over one complete cycle,

$$W = Q_1 - Q_2 \qquad \text{...(5.9)}$$

$$\therefore \qquad \eta = 1 - \left(\frac{Q_2}{Q_1}\right)$$

If $Q_2 = 0$, which means amount of heat rejection is zero, then the engine will have 100% efficiency in converting heat into work. Note that the first law of thermodynamics, i.e. the energy conservation law does not rule out such type of an engine. But some expertise experience assumed that such an ideal engine with $\eta = 1$ which is never possible, even if well suited eliminate with various kinds of losses associated with actual heat engines. It turns out that there is a fundamental limit on the efficiency of a heat engine set by an independent principle of nature, called the second law of thermodynamics.

5.3.3 Second Law of Thermodynamics

The second law of thermodynamics states that *the entropy of any isolated system always increases. Isolated systems spontaneously evolve towards thermal equilibrium—the state of maximum entropy of the system. More simply, the entropy of the universe (the ultimate isolated system, i.e. randomness) only increases and never decreases.*

For example, the second law of thermodynamics (*see* Subsection 1.7.3) can be understood with the rooms of a person, if not gutted and neatened, will habitually become more disordered and unsystematic with time-regardless of how careful one is to keep it clean. When the room is cleaned, its entropy decreases, but the effort to clean it has resulted in an increase in entropy outside the room that exceeds the entropy lost. Simply randomness in the room decides the entropy.

Kelvin-Planck statement: No process is required whose individual result is the absorption of heat from a reservoir and the complete conversion of the heat into work.

Clausius statement: No process is required whose individual result is the transfer of heat from a colder object to a hotter object. It can be proved that the two statements above are completely equivalent.

5.3.3.1 Reversible and Irreversible Processes

If the processes are said to be reversible the changes in the process will back to initial stage, whereas in irreversible process the changes will not back to initial stage. For example, when a water evaporates from its liquid state to gaseous state and it will back to initial stage by its condensation reaction and forms rain. So reversible process is the natural and ideal process. In case if you break a glass into a piece and this will not back to initial stage. This is irreversible process.

Reversible process: In a thermodynamical system a process, the reversible and it will return to initial state of both the system and the surroundings. It means both system and surroundings are returned to their initial states at the end of the reverse process.

Figure 5.4 shows the depiction of system process which undergoes a change from its state 1 to 2. If the system changes thermodynamically from 1 to 2 that are equilibrium with each other. Similarly, from state 2 to 3 and to 4 and finally, it returns to its original state where all the state should be thermodynamically equilibrium with each other.

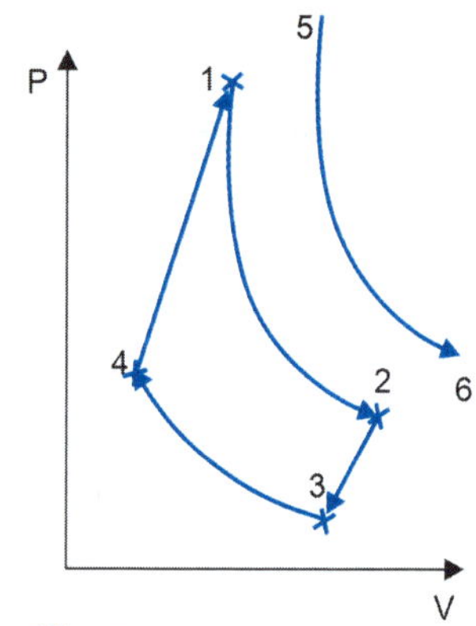

Fig. 5.4: A thermodynamic process—reversible and Irreversible

Irreversible process: From the Fig. 5.4, the system undergoes a change from its state 5 to 6 and it is not back to 5 because of deviated from equilibrium state. In this case, irreversible process occurs as a result of straying away from the curve and not back to initial state. An irreversible process is a thermodynamic process that departs from equilibrium. In terms of pressure and volume, it occurs when the pressure (the volume) of a system changes dramatically and instantaneously that the volume (the pressure) does not have the time to reach equilibrium.

5.3.4 Third Law of Thermodynamics

The third law of thermodynamics states that *the entropy of a system approaches a constant value as the temperature approaches absolute zero*. At absolute zero temperature, the entropy of a system is typically zero and the number of different ground states in a system determined in all cases. For example, the entropy of a pure crystalline substance (perfect order) at absolute zero temperature is zero. This statement holds true, if the perfect crystal has only one state with minimum energy.

5.4 THERMOMETRY

Thermometry in physics is a branch which deals with the measurement of temperature and the device used to measure this temperature is called thermometers. To understand the concept and its technique in thermometry, few basic things that we need to know.

Heat: Heat is one form of energy which is also called thermal energy that flows from a region of higher temperature of body to a region of lower temperature of body. This will occur only when both are in contact with each other.

The process of translational, vibrational and rotational motion effect on heat or thermal energy of a body which is the sum of kinetic energies of all its constituent particles. The SI unit of heat energy is joule (*J*) but in practice the thermal energy can be measure in the unit in calorie.

$\therefore$ 1 cal = 4.18 *J* which mean 1 calorie of heat energy is required to raise the temperature of 1 g of water to raise in temperature of 1 °C.

Temperature: Temperature is defined as the degree of hotness or coldness of the body. A device which is used to measure the temperature, is called a thermometer. It is the property that determines the thermal equilibrium of body with it neighboring system. This thermal equilibrium property can be expressed in a numerical value called temperature.

Highest possible temperature accomplished in laboratory is about 108 K, whereas lowest possible temperature is 108 K. If the temperature close to 0 K is known as cryogenics and if the temperature approaches to very high temperature is called pyromet.

Different Scales of Temperature

Celsius scale: In this scale of temperature, the melting point of ice is considered as 0 °C and the boiling point of water as 100 °C and space between these two points is divided into 100 equal parts

Fahrenheit scale: In this scale of temperature, the melting pointing of ice is considered as 32 °F and the boiling point of water as 211 °F and the space between these two points is divided into 180 equal parts.

Kelvin scale: In this scale of temperature, the melting point of ice is considered as 273 °K and the boiling point of water as 373 K the space between these two points is divided into 100 equal parts.

Relation between different scales of temperatures

$$\frac{C}{100} = \frac{F-32}{180} = \frac{K-273}{100} = \frac{R}{80}$$

Types of Thermometers

1. Gas thermometer
2. Liquid thermometer
3. Magnetic thermometer
4. Radiation thermometer
5. Resistance thermometer
6. Thermoelectric thermometer

5.5 THERMOSTAT

Thermostat is a device which can be used to control and regulate temperature in desired level in different devices. It works on the principle of thermal expansion.

Thermal expansion: It is defined as the alterations of the length, width, height, or volume of any type of material on changing its temperature. It can be easily seen in solids where atoms are tightly packed. It has lot of applications in our day-to-day life. According to this phenomenon substances will be expanded on heating and condensed on cooling.

Working: It works on the general principle of thermal expansion. Based on the expansion or condensation of material governs the switching off or on of the electric circuit. There are different mechanisms of thermostats are in use and the most common types of mechanical thermostats are either bimetallic strips or bellows filled with gas. Whereas the digital thermostats use the same principle, but everything is controlled by a integrated circuits, chips and built-in minicomputer.

Bimetallic strips: Bimetal, the name indicates that the thermostat has two pieces of metals which are varying with its coefficients of expansion but are connected to each other and form a bimetallic strip. This behaves like a bridge to connect or disconnect the electric circuit of the heating or cooling system. When it gets down that causes the heating due to the bimetallic strip conducts current and the system starts like switch on. In this case, the strip gets hot that cause one of the metal pieces to get hotter than the other. The hotter strip expands, causing to break the circuit, the heating or cooling (strip) switches is now off. During this state no current through the bimetallic strip, so the metal starts to cool down gradually which means the expanded metal piece starts condense to its original size. At the end, the heating or colling restarts as shown in Fig. 5.5.

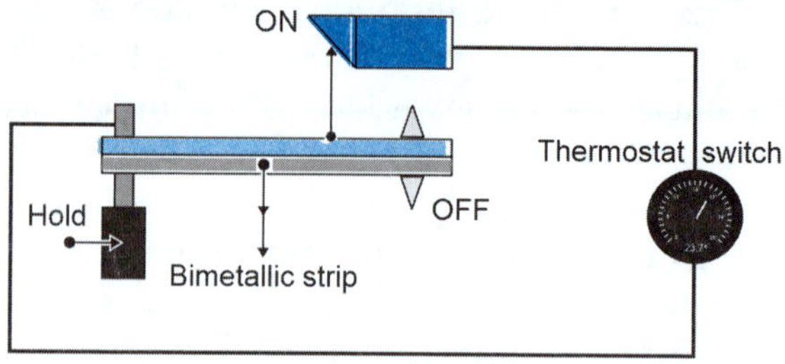

Fig. 5.5: Schematic diagram of thermostat

Advantages

- Easy to handle and not required power source.
- Highly efficient and active device.
- Simple and economy.
- Suitable for a wider range of temperature (<450 °C).

Disadvantages

- Less accurateness.
- Not appreciated in industrials.
- Not suitable for very low temperature.

Types of thermostat

1. Bimetallic strip thermostat
2. Capillary thermostat
3. Digital thermostat
4. Smart thermostat
5. Wax type thermostat

5.6 THERMOCOUPLE

It is a type of device which is used to measure temperature at one point. This thermocouple has a sensor that is used for measuring the temperature in the form of an electric current or the EMF. It consists of two wires of different metals and are soldered together at the ends. The soldered portion is created junction where the temperature is measured. The variation in temperature of the wire induces the voltages and it is recorded with suitable device to display the temperature at one point.

Construction and working of thermocouple: The thermocouple consists of two dissimilar metals. These metals are welded together at the junction points *P* and *Q*. This junction considers as the measuring point. The junction point categorizes into three types.

i. Ungrounded junction
ii. Grounded junction
iii. Exposed junction

Three types of effects are playing a major role in the working principle of the thermocouple.

5.6.1 Seebeck Effect

According to this effect, two different metals are involved where the heat is provided to any one of the metals, the electrons start flowing from hot metal to cold metal. The seebeck effect occurs between two different metals. When the heat provides to any one of the metals, the electrons start flowing from hot metal *P* to cold metal *Q*. Thus induces direct current in the circuit.

Seebeck effect states that *the temperature differences between the two different metals induces the potential differences between them.* This causes flow of electrons between plates producing small voltages.

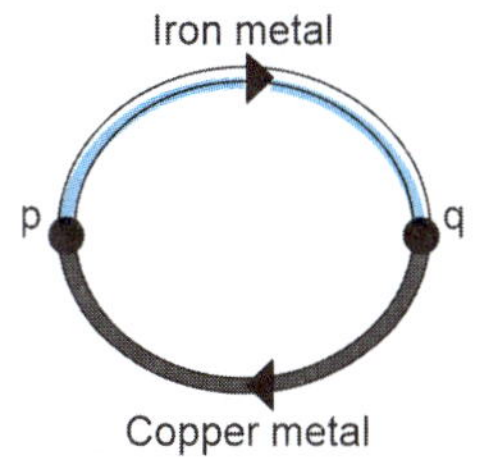

Fig. 5.6: Seebeck effect

5.6.2 Peltier Effect

The Peltier effect is the inverse of the seebeck effect. The Peltier effect states that *the temperature difference can be created between any two different conductors by applying the potential difference between them.*

5.6.3 Thompson Effect

The Thompson effect states that *when two dissimilar metals join together and if they create two junctions then the voltage induces the entire length of the conductor because of the temperature gradient.* The temperature gradient is a physical term which shows the direction and rate of change of temperature at a particular location.

Advantages of Thermocouple

1. The thermocouple is cheaper than the other temperature measuring devices.
2. The thermocouple has the fast response time.
3. It has a wide temperature range.

5.7 CALORIMETRY

Calorimetry is a technique that is used to measure the amount of heat absorbed or released in a reaction. A calorimeter is an instrument used in calorimetry and it is also used to determine heat content, latent heat, specific heat, and other thermal properties of substances. Generally, calorimeter readings are in calories.

Principle of calorimetry: Two bodies of different temperatures are kept together in thermal contact then absorption and release of heat take place. According to second law of thermodynamics states that heat flows from the body at higher temperature to the body at lower temperature which receives heat. This flow of heat will continue until they reach at the thermal equilibrium.

If there is no heat lost during absorption and release of heat in a process, then the amount of heat given up by the body at higher temperature will be equal to the heat gained by the body at lower temperature.

$$\therefore \quad \text{Heat lost} = \text{Heat gained}$$

It is called the principle of calorimetry.

Structure and operation of a basic calorimeter: Considered a chamber consists of two vessel which forms a basic calorimeter. This two-vessel chamber has the outer vessel and the inner vessel. The air is maintained between both vessels that act as an insulator of heat and it is also ensuring that heat is not transferred between the contents of the inner and outer vessels.

The calorimeter is also well equipped with a thermometer and a stirrer in which the thermometer is used to measure the temperature of the liquid in the inner vessel and the stirrer is used to distribute heat equally in the vessel. The calorimeter has an insulating cover with holes for the stirring rod and thermometer.

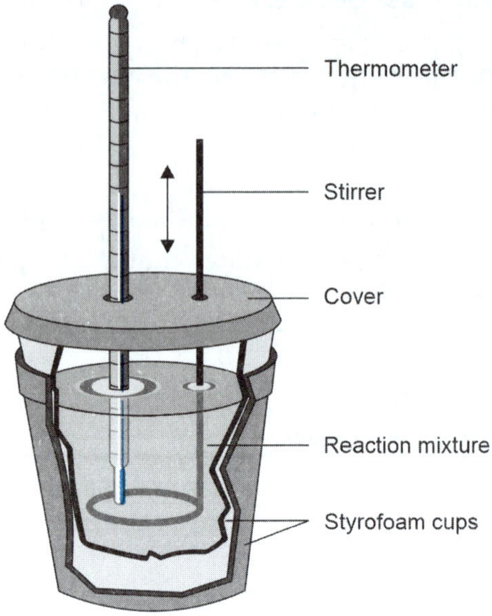

Fig. 5.7: Calorimetry

Working: To measure the specific heat of given material, the inner vessel is now filled with 100 ml water. Later the calorimeter is closed inside the cover for half an hour. At this time, the bulb of the thermometer should be placed in the water. After half an hour, the temperature is read and recorded.

A test tube is half filled with given materials like iron nails or balls. The weight of the material is noted. The filled test tube is now placed in boiling water for 10 minutes. The material is absorb the heat and reach a temperature of 100 °C without getting wet. The calorimeter is opened and the materials are transferred into the inner vessel. The calorimeter is closed immediately after transferred the material.

The temperature of water in the inner vessel is raised because of the heated material and the temperature is noted which is the final temperature of the water and material. The temperature increase of the water and temperature loss of the material is measured.

The temperature increase of water is multiplied by the 100 ml volume of water to calculate the total heat transfer in calories. The total heat transfer is divided by the temperature loss of the iron. Finally, the results are divided by the weight of iron to calculate the specific heat of given material.

Applications: Some of the key application areas of calorimeters are as follows.
- In laboratories like biochemistry and chemistry labs.
- In physics, thermodynamics study.
- In practicals.
- Pharmaceutical and polymer industries.
- Propellant and explosive testing.
- To assess the thermal hazard batteries.
- To study about the thermodynamic properties of different materials like nanomaterial and ceramics.
- To study the liquid crystals properties.

Types of calorimeters: Calorimeters are of many types. Some of the common types are as follows:
- Adiabatic calorimeters.
- Bomb calorimeters (constant volume calorimeters).
- Constant pressure calorimeters.

- Differential scanning calorimeters.
- Reaction calorimeters.

5.8 WATER BATH

It is a type of apparatus that consists of a container filled with heated water. So it can be used to maintain samples in water at a constant temperature for a long period. It is also accommodate with digital or an analogue interface to allow operators to set a required temperature.

It is also used to raise the temperature of reagents, melting of substrates or development of cell cultures. Further, it can be used to enable certain chemical reactions to occur at high temperature. Water bath is a preferred heat source for heating flammable chemicals instead of an open flame to prevent ignition. Different types of water baths are in use based on different applications. 99.9 °C is the ideal temperature for all water baths but in case of temperature above 100 °C, oil bath, silicone bath or sand bath may be used as alternates.

Types of water bath

Circulating water baths: Circulating water bath is used to maintain uniform temperature. It is also called stirrers. The main motive of water bath is ideal for temperature uniformity. Circulator is used to regulate the temperature of water at specific level. In enzymatic and serologic experiments, this type of water bath is very essential and consistency is critical.

Non-circulating water bath: This is the convection type of water bath which is adapted primarily, later it is derelict because water in the water bath is heated uniformly instead of circulating. Therefore, it is less accurate in terms of temperature control. This is considered as disadvantage in this type of water bath and it is modified with stirring to non-circulating water baths to create more uniform heat transfer.

Shaking water bath: The name of the water bath itself implies it function. It has an extra control for shaking, that provides liquids to move around. Even this can be turned on or off. In microbiological practices, constant shaking allows liquid-grown cell cultures grown to constantly mix with the air.

Precautions

- It is moisture sensitive or pyrophoric reactions. Do not heat a bath fluid above its flash point.
- It should be monitored regularly and have to use only distilled water so that it can prevent salts to deposite on the heater.
- Disinfectants can be added to prevent growth of organisms.
- For decontamination process, we have to raise the temperature to 90 °C or higher.
- It should be operated in well ventilated area
- To prevent evaporation and raising temperature, it is always closed.
- Set up on a steady surface away from flammable materials.

5.9 INCUBATOR

An incubator is a device used to grow and maintain microbiological cultures or cell cultures. The main purpose of the incubator is to maintain optimal temperature, humidity and carbon dioxide (CO_2) level and oxygen content inside. It has a vital role in experimental work of cell biology, microbiology, molecular biology, and biotechnology.

Louis Pasteur, a French chemist, and microbiologist who investigate his research on microbial fermentation. He used the small opening underneath his staircase as an incubator for his research. It is often used in the poultry industry to act as a substitute for hens. Nowadays, various brands of incubators are commercially available.

The simplest incubators as shown in Fig. 5.8, are insulated boxes with an adjustable heater, typically going up to 60 °C to 65 °C (140 °C to 150 °F), though some can go slightly higher (generally to no more than 100 °C). The most used temperature bath for bacterial culture such as *E. coli* and other mammalian cells is approximately 37 °C (99 °F), as these organisms grow well under such conditions. For other organisms used in biological experiments, such as the budding yeast, *Saccharomyces cerevisiae*, a growth temperature of 30 °C (86 °F) is optimal.

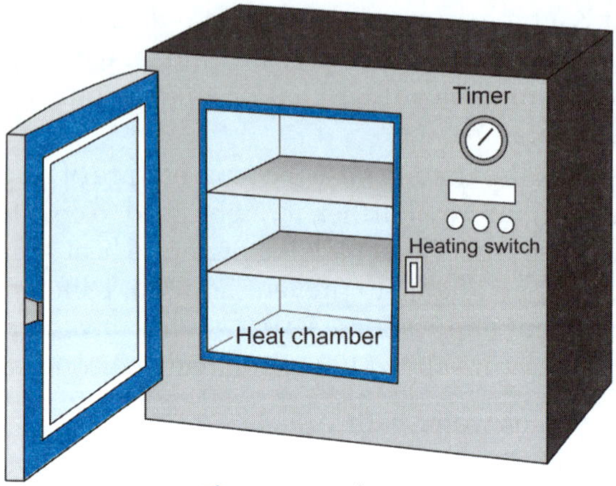

Fig. 5.8: Incubator

More elaborate incubators can also include the ability to lower the temperature (via refrigeration), or the ability to control humidity or CO_2 levels. This is important in the cultivation of mammalian cells, where the relative humidity is typically >80% to prevent evaporation and a slightly acidic pH is achieved by maintaining a CO_2 level of 5%.

Applications

- Breeding of insects and hatching of eggs in zoology
- Controlled sample storage
- Growing cell cultures
- Growing of crystals/protein crystals
- Reproduction of germ colonies and subsequent determination of biochemical oxygen demand (wastewater monitoring)
- Reproduction of germ colonies with subsequent germ count in the food industry
- Reproduction of micro-organisms such as bacteria, fungi, yeast or viruses
- It is also used in other fields like genetic engineering, medical treatments, tissue culture, pharmaceutical studies, hematological studies, biochemical studies, food processing, cell aeration, plant and animal studies, solubility studies, fermentation studies and bacterial culturing.

5.10 AUTOCLAVE

Autoclave is a process of sterilization with the help of steam which is under reduced pressure and temperature above 100 °C. This steam sterilization process is carried out in the chamber called an autoclave.

Construction: This autoclave consists of chamber which is made up of stainless steel and it is also provided with moist heating for sterilization. It has a covered with steam vent, a pressure, and a safety value. To make an air tight inside the chamber, it is provided with rubber gasket. The heating is done in the chamber by an electrical heather, that is, fitted at the bottom. So, it can convert the water to the steam. The steam is supplied to the autoclave chamber where the perforated inner chamber is placed on the stand. The material is sterilized which is loosely packed into it as show in Fig. 5.9.

Principle: The principle of the autoclave is when the vapor pressure of water is equal to that of surrounding atmosphere then the water starts boil. If the pressure inside the closed vessel increases, the temperature of water boils also increases. This is a better condition to sterilize because the saturated steam has better penetrating power.

Fig. 5.9: Autoclave

Steam condensed to water and gives up to latent heat to that surface when it comes in contact with the cooler surface. The water of the condensation is ensuring moister conditions for killing of the exposed micro-oraganisms. For example, bacterium is susceptible and this moist heat to bacterial proteins coagulates rapidly and condensed water ensures moist conditions for killing the microbes present.

Working

- Once perforated plate in removed from the chamber, a sufficient amount of water is poured into the chamber. The level of the water is adjusted to avoid the contact between water and perforated plate.
- The lid of autoclave is closed with the wing nuts and bolts. Later the autoclave is a switched onto heat the water. The vent is now opened and safety valve is adjusted to ensure the required pressure.
- Whem steam starts coming out from the vent and it continuous for 5 mins and it is then closed. It indicates that air has been removed. The steam pressure starts raising and it comes to the desired pressure.
- Therefore, 10 lbs per square inch with corresponding temperature 121 °C after the stated period, switch of the autoclave. Allow it to cool to about 40 °C before opening the vent. When whole of the steam inside the autoclave is removed, the lid is opened and the sterilized material is removed.

Type of Autoclave

1. Downward displacement laboratory autoclave
2. High security autoclave
3. Large simple autoclave
4. Low temperature steam
5. Media preparators

6. Multipurpose laboratory autoclave
7. Porous load sterilizers
8. Simple laboratory autoclave
9. Transportable bench top autoclaves

Advantages

1. Autoclave has steam to provide intense heat and moisture for decontaminate objects.
2. It can destroy micro-organisms more efficiently than dry heat
3. In autoclave, the materials are exposed to a lower temperature for a shorter period only.
4. It is used for sterilization of many official injections.
5. Equipment or parts of rubber or plastics which can be withstand the temperature can be sterilized easily.
6. A large quantity of materials can be sterilize.

Disadvantages

1. It is not suited powders and oils.
2. It cannot be used to sterilized of material that not withstand the temperature of 115 °C–116 °C.

5.11 REFRIGERATORS

A refrigerator is type of enclosure that interior temperature is kept lower than surrounding environment. The term "refrigerator" was coined by a Maryland engineer, Thomas Moore, in 1800. Moore's device would now be called an "ice box"

It is worked by passing a cool refrigerant gas around food items that is placed inside and the refrigerant can absorb heat from them and then loses that heat to the relatively cooler surroundings on the outside of the refrigerator.

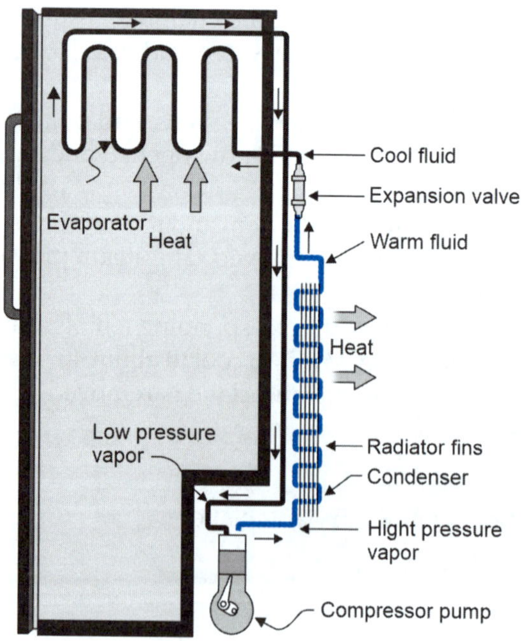

Fig. 5.10: Refrigerator

Refrigerator working principle: The working principle of a refrigerator is very simple. In general, it removes heat from one place and its deposition to another. When you pass a low-temperature liquid close to objects. Heat from those objects is transferred to the liquid, which evaporates and takes away the heat in the process. The gas called refrigerant tends to become hot when it is compressed and become cold when it is expanded, which helps a refrigerator cool the stuff being kept inside always.

Parts of a refrigerator: The following are the parts of a refrigerator and consist of a few key components that play a vital role in the refrigeration process:

Expansion valve: It is also called the flow control device. The flow of liquid refrigerant can be controlled by the expansion valve. Actually, this device is a very small device and it is very sensitive to temperature changes of the refrigerant.

Compressor: The compressor consists of a motor that 'sucks in' the refrigerant from the evaporator and compresses it in a cylinder to make a hot, high-pressure gas.

Evaporator: Evaporator is one of the important parts in refrigerator. It has crooked tubes that helps to absorb heat and blown it out through a coil. It absorbs heat not only from fridge but also from the fold stuff and absorbed heat turns liquid refrigerant into vapor. This will be back to liquid state after release the heat from it.

Condenser coils: The condenser placed outside of refrigerator with external fins or crooked tubes. It helps the liquefication of vaporized refrigerant by releasing the heat to the surroundings. Once the temperature drops then refrigerant tends to condensation and become liquid.

Refrigerant: It is a type of specialized chemical that has ability to become gas by receiving heat and become liquid by release heat. CFCs are commonly called Freon and it is the brand name of R12 which is commonly used as refrigerant. This was later replaced with HCFC called hydrochlorofluorocarbon. The brand name of HCFC is R22. At present various refrigerants/coolants are commercially available now. They are R410A, R22, R32, R290, R134A, R600A. This can be used for both air conditioners and refrigerators

Types of Refrigerators: Depending upon the purpose of usage and application the refrigerator, it can be classified as follows:

1. Commercial refrigeration.
2. Domestic refrigeration.
3. Industrial refrigeration.
4. Marine and transportation refrigeration.

Chapter

6

Review of Physics—Instrumentation

6.1 PHYSICAL BALANCE

A physical balance is also called beam balance. It is a device that used to measure the mass of an object. It has two pans with equal weights which are balanced by a beam and the known weight should keep on one side whereas the unknown weight to be measured should keep on the other side of balance.

Working Principle: A beam balance works on the principle of moments. According to the principle of moments at equilibrium, the anti-clockwise moment due to the weight of an object on left pan of the beam is equal to the clockwise moment due to the standard weights on the right pan of the beam.

Construction

- A metallic pointer and an ivory scale are supported by the pillar.
- The plumb-line attached with pointer beam indicates the position of the balance in horizontal.

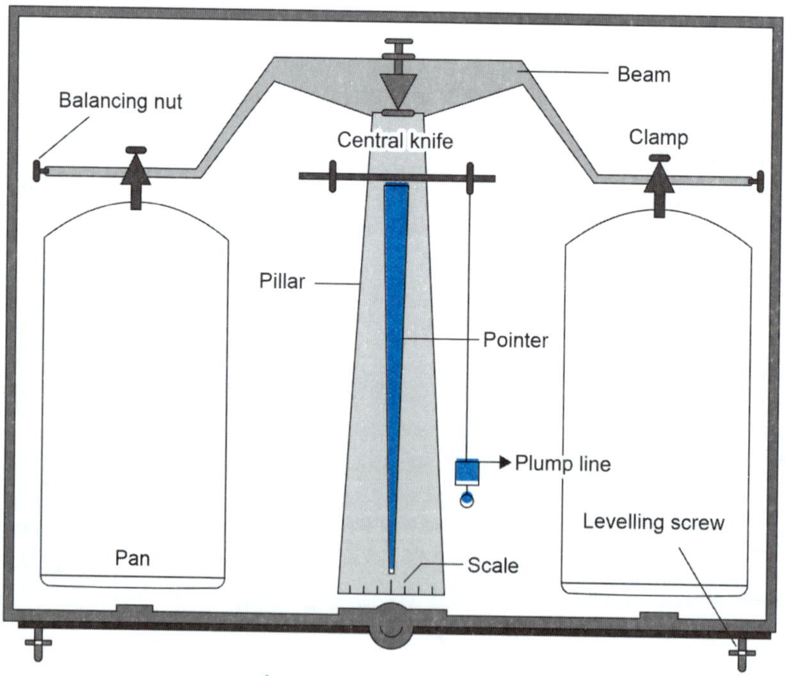

Fig. 6.1: Physical balance

- Initial adjustment made in balance that the pointer remains on zero mark on the ivory scale before the balance to start to measure
- To avoid the effect of surrounding air, the device is enclosed in a glass box.
- A weight box which is provided with balance has standard weights with spatula/ forceps.

Use of Beam Balance

- It is used to measure the mass of unknown object with known standard masses.
- Unknown mass is kept in one pan whereas known masses are kept in another pan such that the pointer should lie at the centre that determine the exact mass.
- To maintain the pointer at the center subsequent mass is added. After the point at center in rest state, the total known mass in the other pan is calculated, which gives the mass of the object.

Mandatory for Good Balance

- For a good conditioned balance, the beam remains horizontal when equal weights or no weights are placed on two pans. It can be achieved by adjusting the screws of balance to maintain the arms of the two balances are of equal length, the weights of two pans should be equal. The centre of gravity of the pan beam should pass through the fulcrum, when the beam is horizontal.
- The sensitivity of the balance is high so that it can measure the weight of very small object that shows large deflection in the scale. It can be achieved easily by increasing the length of the arms of the beam.
- While measuring the weight of an object, a beam balance should be stable. It can be achieved by making the beam heavy.

Sensitivity of a balance: Sensitivity of the balance can be identified by placing a very small object to measure weight accurately. A balance is said to be sensitive when it can register small differences in weight. The physical balance beam is most sensitive where its center of gravity lies slightly below its rotational axis. If the center lies too far below the suspension point, the balance is too stable and, therefore, has low sensitivity.

Precautions in using a Beam Balance

- The weights from the weight box should be carried with the help of forceps. This is because to avoid the change in its weight due to the presence of moisture and dust particles from hands.
- The box and its weights should be clean and dry.
- When the pointer is near zero mark, try the weights in the descending order.
- The beam should be lowered each time before adding or removing weights from the pan.
- It should be handled gently to avoid knife edges from breaking.

6.2 CHEMICAL BALANCE

Chemical balance is a type of delicate instrument, that is, used to measure the weight of small amounts of substances accurately. The accuracy of an instrument can be determined by its sensibility of measuring a tiny object up to the fourth place of decimal and even can recognize a minimal deviation of tiny object. The chemical balance also has a vital role in volumetric analysis as the substances, that is, used to prepare standard solutions are weighed. The most widely used types of balance are balances with a capacity of 100 g and a sensitivity of 0.1 mg.

Principle: The working principle of chemical balance is an electromagnetic force restoration (EMFR). The beam balance is comparable to the weighing balance principle. One side weight is laid known as coil arm and the coil arm is attached to the other arm of the weighing balance. In this system, the force exerted by the object being weighed is lifted by electromagnet. It tries to move out of the electromagnetic force of the magnetic field. This is the basis of the rapid and accurate measurement of the weighing balance. High current is passed through the coil with the help of the electrical regulator. This high voltage current helps in keeping the weighing balance in a neutral position.

Weighing Balance Uses: This type of balance is widely used in the pharmaceutical, chemical and food industry. Due to their high accuracy weighing balance is used in various laboratories like biology laboratory, chemistry laboratories, physics laboratories, etc. Accuracy is the major role in the chemical process, so that the balance has a significant role in chemical reactions and research.

Different Types of Balance: The different types of balancing instruments are as follows:

- Analytical balance
- Chemical balance
- Beam balance
- Physical balance

Analytical balance: Analytical balance is a type of balance which is used to measure a small number or quantity of substances (generally chemical). It is present in the digital form and handling of an analytical balance is user-friendly that means very easy to use.

Beam balance: Beam balance is an instrument, used to measure the mass of the substance under the action of the gravitational force. It is generally used in physics laboratories.

Chemical balance: Chemical balance is a type of beam balance equipment. It is used in different types of laboratories like chemistry lab, physics lab, etc. It works on the principle of electromotive force restoration (*see* Section 6.2).

Physical balance: Physical balance is an instrument used to measure the gravitational mass of substances (*see* Section 6.1).

6.3 PROPERTIES OF MATTER

Matter in physics is anything that has mass and volume which includes all atoms and all subatomic particles. All mixtures of compounds, all objects that we encounter around us also a type of matter. In addition to it understanding states of matter, phase changes and chemical properties of matter underlie much of physics. In order to understand in detail about matters, it is important to understand physical quantities like mass, density and pressure.

The properties of matter include physical and chemical properties that can be measured, such as an object's color, density, hardness, length, malleability, mass, melting point, odor, temperature, volume, etc. There are some properties which are unchanging, like density of gold, while other properties depend on quantity of matter like the mass of a golf ball compared to the mass of a cricket ball or the properties of the combination of materials.

Elements of all atoms have a special character, but some properties of an element will be the same and at the same time some element properties can be varied drastically through chemical reactions. For example, pure potassium is a soft metal that will react with oxygen to produce superoxide and with water forms KOH, hydrogen, and gas. Even heat can be produced through hydrogen by ignition. In other case, an ice block changes its physical properties when it melts and at the same time chemically it is the same water in either state. This state is reversible when you remove heat from the water. Depend on the physical and chemical properties, materials are varied substantially.

Physical Properties

- Boiling point
- Colour
- Density
- Mass
- Melting point
- Shape
- Size
- Volume

From the above physical properties, all material have some physical properties that are different from each other. These physical properties can be changed, sometimes permanently and sometimes reversibly. For example, when ice melts become water and it can be refrozen when the water loose its heats, but when the wood is burned many of its atoms leave from wood as gas and it is irreversible.

Chemical Properties: Chemical properties of matters depend on a substance and its reactions on other substances. For example, when a cup of water is mixed with vinegar that solution helps to separate part of baking soda. Chemical properties can be observed in this reaction, that is, carbon dioxide begins to bubble up from the baking soda is mixed with vinegar but not from the portion to which water had been added. During a chemical reaction a new substance is formed with new and different properties than its original substances.

All properties of matters are either physical or chemical properties. Physical properties are either intensive or extensive. Extensive properties, such as mass and volume, depend on the amount of matters. Intensive properties, such as density and color, do not depend on the amount of matters. Physical properties are measured without changing its chemical identity whereas chemical properties can be measured only when chemical identity is changed.

There are different states of matters describe the distinct forms that are solid, liquid, gas, or plasma. Even more than these four states of matters are also available, but we will discuss only occur in extreme situations that are not normally encountered on earth.

Solid: Solids can be considered as a rock which is hard. Solids are hard because it has a definite size, shape and volume. It has a definite shape because the molecules of solid materials are tightly packed together. In a solid, particles are unable to move due to its tightly packed material and these particles have very low kinetic energy with high potential energy. All electrons of each atom are in constant motion which causes atoms with small vibration and at the same time atoms are fixed in their position. If atom cooled to absolute zero, i.e. 0 K, all motion will stop. Increasing pressure will not compress the solid instead it remove the space between part that make solid become smaller in volume.

Liquid: A liquid is a type of different states of material that can change its shape when it poured into container and it can feel wet. Examples for liquid states are water, blood, gasoline, and lemonade which are liquid at room temperature. If a different types of molecules dissolved in a liquid which forms a solution in liquid form. Lemonade is a solution of water, sugar, and lemon; a combination of molecules. In this phase, the substance has more kinetic energy than those in a solid. The liquid particles are in irregular arrangement, but are still very close to each other so it can also have a definite volume. Liquids cannot be compressed like solid because particles of a liquid flow around each other, so it may not have definite shape.

Gas: Unlike solid and liquid, gas molecules are more spread out and disorganized. Gas molecules can be filled up any container regardless of size or shape of container. Gas particles have high kinetic energy, when it is unconfined, the particles of a gas will spread out indefinitely and when it is confined, the gas will be completely filled up in the container.

When gas molecules are put under pressure by reducing the volume of the container ($PV=nRT$) or pressurized gas into a container that reduce the space between particles. If the 'V' kept constant and the temperature of the gas increases, then the pressure will also increase. Conversely, if the pressure is decreased, the temperature will also decrease. If gas molecules are compressed that forms a liquid in the container and when it is released from the container, it changes back to a gas. So, gas molecules have enough kinetic energy to overwhelmed intermolecular forces which hold solids and liquids together, therefore, a gas has no definite volume and no definite shape.

Plasma: A plasma is a fourth type of state of matter which also a superheated fundamental state of matter. Plasma matter consisting of equal numbers of positively charged ions and negatively charged electrons. The characteristics properties of plasmas are significantly differ from ordinary neutral gases so that plasmas are considered a distinct "fourth state of matter." It is not a commonly available matter on earth, but is a very common state of matter. For example, the noble gases like helium, neon, argon, krypton, xenon, and radon are often used to make lighting signs by using electricity to ionize it to form plasma state.

6.4 DIFFUSION

In physics, diffusion is the movement of particles, ions, molecules, atoms or substance from the region of higher concentration to the region of lower concentration. The movement will occur until the process reaches an equilibrium state. A unique feature of diffusion is the process at the end mixing or mass transport occurs without bulk motion. Thus, diffusion should not be confused with bulk motion like convection (movement of a fluid where density gradient plays a role) or advection (movement of material where velocity plays a role), which are other transport mechanisms that use bulk motion to move particles from one place to another.

Molecular diffusion is often called simple diffusion. Once the temperatures reaches above absolute zero the thermal motion of all particles involves in diffusion. The rate of the movement is mainly depending on temperature, viscosity of the fluid and the size of the particles. The resultant diffusion is gradual mixing of materials. Diffusion is of two types, namely simple diffusion and facilitated diffusion.

6.4.1 Simple Diffusion

Simple diffusion denotes the process where the movement of molecule or substance (chemical or electrochemical) from the region of higher concentration to the region of lower concentration until it become saturate or equilibrium. This process happens mainly in gases, liquids and solutions. For example, diffusion process in human body is the transport of oxygen from lungs (high concentration) into the blood (low concentration) and similarly the transport of carbon dioxide from blood to the lungs towards concentration gradient.

In addition, a substances can also diffuse over the semipermeable membrane if the membrane is permeable to diffuse substance. The movement of nonionic substances, like oxygen, carbon dioxide and alcohol, which are lipid soluble through the lipid bilayer of plasma membrane. The membrane made up of lipids, proteins and carbohydrates which are arranged like layers of plasma membrane. The electrolytes can diffuse through these layers.

6.4.2 Facilitated Diffusion

This is type of passive process in which large molecules are unable to passes through the semipermeable membrane, e.g. amino acids and glucose. These types of substances can pass the membrane with the help of mediator protein called facilitation diffusion.

In this mechanism, the substance or molecule can bind to a mediator protein called specific transporter which presents on the cell membrane. After binding of the substance, the transporter now it changes its shape and releases the solute on the other side of the membrane. In this way, the solute is transported from the region of higher concentration to the region of lower concentration along the concentration gradient. The rate of facilitated diffusion depends on the concentration gradient of substance and the number of transporters.

6.5 OSMOSIS

There is a difference between diffusion and osmosis, in diffusion the movement of solute from the higher concentration region to the lower concentration region whereas in osmosis the movement of solvent from the lower concentration region to the higher concentration region. In both process the movement happen till it reach saturation condition.

The process of osmosis is explained as follows, if two solutions of different concentrations are separated through a semipermeable membrane where the membrane allows only a solvent not a solute. The diffusion of solvent from the lower concentration region to the higher concentration region is called osmosis.

The process has a great importance in biology, the transport of water and other molecules crossing the membrane has an important role in living organisms. It has two types, (i) endosmosis and (ii) exosmosis. This classification is mainly differed only in the direction of solvent movement. In endosmosis, the water or solvent moves inside the cells, during this process the cell starts swelling. In case of exosmosis the water or solvent flows out of the cells, during this process the cell starts shrinking.

6.6 CENTRIPETAL FORCE AND CENTRIFUGAL FORCE

Moving a thing necessitates the use of force which means the force is responsible for any moving objects. Depending on the type of motion an object the force which acts on it also operates differently. When the objects move in curvilinear motion, two types of forces come into play.

i. **Centripetal force:** The centripetal force is the force that acts on the object to operate in the direction of circular path. The centripetal force is always perpendicular to the direction in which the object is being displaced. When an object accelerates in response to it changes in velocity, it might modify its speed or direction of travel and the object is directed toward the axis of rotation or centre of curvature. To understand the concept in a better way, if a moving item in a circular path is constantly changing its direction, which means it is accelerating. If an object is moving on a circular path, according to Newton's second rule of motion states that the centripetal force of an object moving in a circular path always acts towards the centre of the circle. The unit of centripetal force is Newton.

   Examples:

   1. If a car makes a turn, the force of friction acting on the wheels which turned and the car provides centripetal force to attain circular motion.
   2. If a bucket of water is tied up with a string and rolled in a circle, the force acting on the bucket provides centripetal force to attain circular motion.

ii. **Centrifugal force:** Centrifugal force is a type of inertial force that act on a body which is moving in a circular path and experiences an outward force that is away from the centre. It arises from the body's inertia and appears to act on a body that is moving in a circular path. The force depends on the mass of the object and the distance from the centre of the circle and also the speed of the rotation. If the object has more mass the force of the movement and the speed of the object will be greater. If the distance is far from the centre of the circle, the force of the movement will be more. The centrifugal force drives the object away from the centre. It is a fictitious force.

Table 6.1: Differences between centripetal force and centrifugal force

Centripetal Force	Centrifugal Force
If the object travels at a uniform speed in a circular path, it is called centripetal force	If an object moving in a circle and experiences an outward force and it is called the centrifugal force
The object has the direction along the centre of the circle from the object approaching the centre	The object has the direction along the centre of the circle from the centre approaching the object
A satellite orbiting a planet is an example of the centripetal force	Mud flying of a tire is one example of the centrifugal force

6.7 CENTRIFUGE

Centrifuge is a technique for separating substance and it is designed to accelerate the sedimentation process by using centrifugal force. The particles or substance separated from the solution based on their size, shape, density, and viscosity of medium.

Principle of Centrifugation: Particles can separate from the solution depends on following conditions:

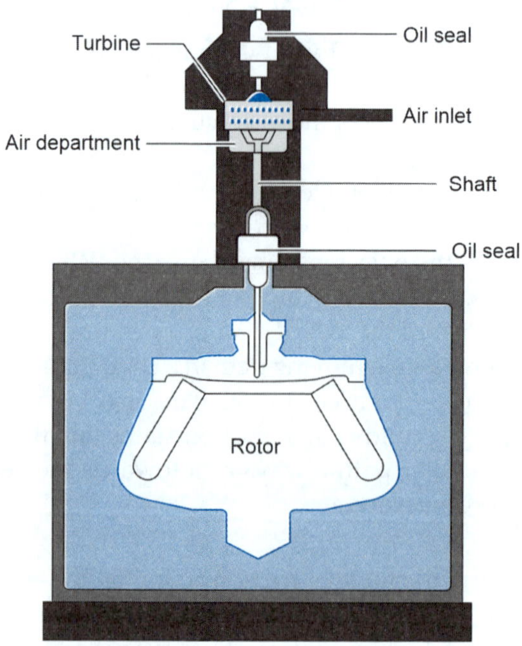

Fig. 6.2: Centrifuge

- Particles with high density can sediment at the bottom of solvent whereas low density particles are floats at the top.
- Particles with high density will move faster. In case of isopycnic (no density difference) the particles will be at same position
- In case of tiny difference in density to separate particles in solution, gravity plays a major role instead of "centrifugal force"
- Centrifugation is done by applying a strong perpendicular force to the axis of spin which process particles outward.
- Centrifuge works on the sedimentation principle, where the centripetal force causes the separation of particles and substance move outward.

The two general types of centrifuges used in the laboratories are:

a. The swing out head type (horizontal rotor), and
b. The angle head type (fixed angle rotor). A rotor is the device used to hold centrifuge cups and tubes.

In the case of swing out head centrifuge, when the centrifuge is at rest, the tubes are in a vertical position and during the process the tubes are in horizontal position. The speed of rotation is more than 5000 rpm. But practically it has some limitations, in case of swing out head type centrifuges, the excess heat developed due to friction of the tubes with air. So that the angle head of centrifuge tubes is positioned at an angle to the vertical axis of rotation as shown in Fig. 6.2. Due to the very low friction of air, angle head rotor units can attain much higher speeds. The centrifuges also capable of producing speeds up to 100,000 rpm which are generally referred to as ultracentrifuges.

The term rpm (revolutions per minute) gives only the centrifugation speed and does not express the centrifugal force. A better expression of forces generated by centrifuge is RCF (relative centrifugal force) which is calculated by considering the speed of rotation (rpm), and the radius from the center of rotation. RCF can be calculated with the value of rpm by using the following formula

$$\text{RCF} = 11.2 \times r\,(\text{rpm}/1000)^2 \text{ or RCF} = 1.12 \times r \times 10^{-5}\,(\text{rpm})^2$$

Where r is the radius of the rotor (cm)

For example, 1,500 rpm and 1,500 RCF are two different things. RCF is a function of both rpm and the rotating radius of the rotor (R). Therefore, RCF is a quantitative number while rpm is only relative number when computing how much sedimentation force is being applied to a sample.

Important Components of a Centrifuge

- The centrifuge head (rotor) with tubes or cups with cushions (rubber pads) at the bottom inside the cups or tubes
- Motor drive assembly (a commutator type driving motor)
- The chamber which encloses the above-mentioned internal parts
- A power switches
- Speed control and the additional parts
- A timer
- A tachometer (to read rpm of the machine)
- Graphite brushes

Uses: The centrifuge is used in a laboratory for various purposes such as

- The separation of serum or plasma from red blood cells
- Separation of sediment in urine
- Separation of protein free filtrate
- Washing of red blood cells by normal saline
- Separation of antigen bound fraction or antibody bound fraction from the free fraction in immunoassays.

Working of a Centrifuge: The centrifuge motors are supplied with DC motors which is connected in series, when voltage is increased that turn rotor faster or if it supplied with AC motors in which speed adjustment is achieved through stepwise reduction of the number of poles in the magnetic field. Electrical contact to the rotor is provided by graphite brushes. The speed is controlled by a potentiometer of the centrifuge, which raises and lowers the voltage supplied to the motor.

Care and Maintenance

- Always make sure that the cover lid is closed while the centrifuge is operating.
- Before the centrifugation, the centrifuge tubes and the cups should be balanced properly. Run the centrifuge with full accessories and units like rotor, head, tube, etc.
- Observe for usual normal vibration noise during operation, if the tubes inside are not balanced properly. This should be checked immediately.
- Place the centrifuge on a firm base. It should not be placed near a sensitive equipment like photometer or spectrophotometer. It causes electrical noise and high current drain.
- The chamber should be kept clean since it may contain biohazardous materials (chemical, microbiological or radioactive).
- The rotor will not function, if the graphite piece wears out completely. In that case, it is necessary to replace it with a new graphite brush.
- The working of centrifuge should not be done near any combustible fluid.
- Cover the centrifuge, when not in use.

Functional Verifications

- **RPM calibration:** The speed of centrifugation should be checked by using an external device for good accuracy. Values obtained with the external measuring device should agree within 5% of those from a built-in tachometer.
- **Timer:** The timing intervals of the timer should be checked against an accurate stopwatch or electronic timer.
- **Temperature:** In the case of a refrigerated centrifuge, the temperature of the chamber should be checked against a certified thermometer.
- Functions of the centrifuges should be verified after every six months.

Preventive Maintenance

- According to the manufacturer's direction lubrication of bearings, motor components, replacement of graphite brushes, current leaking and checking of fuse should be done.
- The line cord, plug, lamps and wiring should be examined at regular intervals.
- The rotor, cups, tubes and control knobs should be examined frequently.

6.8 pH METER

A pH meter provides a value of liquid level is to determine acidic or alkaline nature. pH value of any liquid is mainly depending on the concentration of hydrogen ions. Acids dissolve in water and produce positively charged hydrogen ions (H^+). The higher concentration of hydrogen ions resulted the stronger acid nature of liquid. Similarly, alkali or bases dissolve in water produces negatively charged hydrogen ions (OH^-). The stronger base is determined by the higher concentration of negatively charged hydrogen ions.

If the pH value of any liquid is 7 which indicates a neutral solution. In case of pure water pH value is 7. If pH value is less than 7 indicate an acidic solution while a pH value is greater than 7 indicate an alkaline solution. A solution with pH value very near to 1 is highly acidic and a solution of pH value nearer to 14 is highly alkaline.

Basic Principle of a pH Meter

- A pH meter is associated with a probe, which is accommodate with two electrodes. This probe provides an electrical signal to a meter which displays reading in pH units. The glass probe has two electrodes in which one is a reference electrode and the other one is made up of glass sensor electrode.
- Both electrodes have a hollow space that filled with potassium chloride solution and suspended with a silver chloride wire into it. The glass sensing electrode has coated with silica and metal salts. It measures the pH of solution by measuring hydrogen ions concentration which surrounds the tip of the thin-walled glass bulb. The reference electrode has a bulb made up of a non-conductive glass or plastic.
- If one metal electrode is in contact with liquid and the other reference electrode is in solution that produces a voltage difference due to electron mobility. A pH meter measures potential between a known liquid and an unknown measuring liquid outside. The glass bulb allows hydrogen ions to interact with the glass to measure potential of hydrogen ions. To complete the electrical circuit, a reference electrode is also needed.

Silver Chloride Electrode: A silver chloride electrode is a type of reference electrode which is commonly used in electrochemical measurements. For example, it is usually the internal reference electrode in pH meters. The reaction is mainly in between the silver metal (Ag) and its salt—silver chloride (AgCl, also called silver (I) chloride).

The corresponding equations can be presented as follows:

$$Ag^+ + e^- = Ag(s)$$

$$AgCl(s) + e^- = Ag(s) + Cl^-$$

This reaction is characterized by fast electrode kinetics, meaning that a sufficiently high current can be passed through the electrode with the 100% efficiency of the redox reaction (dissolution of the metal or cathodic deposition of the silver-ions). The reaction has been proved to obey these equations in solutions of pH values between 0 and 13.5.

Working of pH Meter: The pH meter measures the potential difference between the liquids and its changes across the membrane. The potential difference must be obtained between two places; one is the electrode that contacts the internal solution. A second place is obtained by connecting to a reference electrode, which is immersed in the steadied (unknown) solution as shown in Fig. 6.3.

The overall working principle of pH sensor and pH meter depends upon the exchange of ions between sample solution (pH 7 buffer) and studies solution. The porosity of the glass membrane decreases with the continuous use that decreases the performance of the probe.

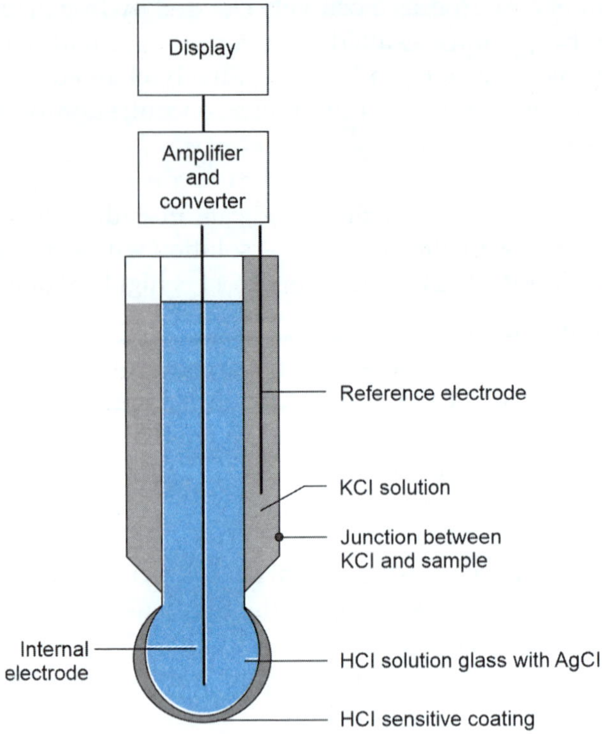

Fig. 6.3: Glass electrode pH meter

Types of pH meters: Nowadays pH meters are developed in advance level and it ranges from simple and inexpensive pen-like devices to complex and expensive laboratory instruments with computer interfaces. The output can be digital or analog, and the devices can be battery-powered or rely on line power. Some versions use telemetry to connect the electrodes to the voltmeter display device.

Specialty, meters and probes are also available for use in some special conditions, such as harsh environment and biological microenvironments. There are also holographic pH sensors, which allow pH measurement calorimetrically, making use of the variety of pH indicators that are available. Additionally, there are commercially available pH meters based on solid electrodes, rather than conventional glass electrodes.

- Cheese manufacture
- In agricultural—soil evaluation
- In preparing wine
- In the swimming pool
- Production of soaps

Chapter

7

Sound

7.1 SOUND AND NOISE

In physics, sound is a mechanical vibration from equilibrium states that propagates as an acoustic wave. According to physics when objects or things vibrate that produce sound, which is a kind of energy. It can travel through solid, liquid, and gaseous states as an auditory wave. The ideal audible sound frequency ranges from 20 Hz to 20 kHz. Ultrasound is a sound with a frequency greater than 20 kHz which is beyond the audible range.

The sound what we hear is the effect created by to-and-fro motion of particles in any medium. Back-and-forth motion is generally called vibration. Sound waves travels through a medium by contracting and expanding portions of the medium. Expansion and compression that created by a little pressure difference is the sound, that is, what we hear. As a result, it is a mechanical pressure and displacement wave.

There are many kinds of sounds, based on the characteristics like pitch, loudness, amplitude, and frequency of the sound wave. In these characters are not all pleasing and pleasuring our ears. The sound is divided into two categories—noise and melody, depending on whether we enjoy it or not. Music is what we satisfy our auditory senses. It is determined by a variety of circumstances and it may differ from person to person. The definition of noise, on the other hand, is fuzzy and unclear. The distinction between musical sound and noise is hazy. What is music to one person may be noise to another.

7.2 SOUND WAVES

Generally, waves are like repeating pattern similarly a sound wave is frequently referred as a pressure wave and it consists of a repeating pattern of high-pressure and low-pressure areas travelling through a medium. If a detector is placed to detect a sound wave, whether it is normal to human ear or a man-made instrument. The instrument is principled to detect by measuring pressure changes when the sound wave interacts with detecting device. The detector would detect a high pressure at one point in time, which would correlate to the arrival of compression. The detector may also detect normal pressure at a later point in time. Finally, a low pressure would be observed, indicating that a rarefaction had arrived at the detector site. So, the pressure changes are measured by the name of compression and rarefaction.

Sound is a mechanical wave that created by the back-and-forth vibrating of particles. When a sound wave travels through a medium from left to right through air, the energy of the sound wave causes particles of air to be moved both rightward and leftward. The particle motion is parallel (and anti-parallel) to the energy transmission direction, so because of this sound waves in the air are known as longitudinal waves.

7.2.1 Frequency, Wavelength and Wave Speed

Frequency: Frequency defines the number of oscillations or waves that pass in a fixed place at a given amount of time. In a simple way, it is defined as number of cycles per second. For example, if 10 waves occupy in one second, then the frequency is 10 Hz as shown in Fig. 7.1a. If the number of frequencies is more in 1 sec, then it is called high frequency radio waves. Similarly, the number of frequencies are less in 1 sec, then it is called low frequency radio waves as shown in Fig. 7.1b.

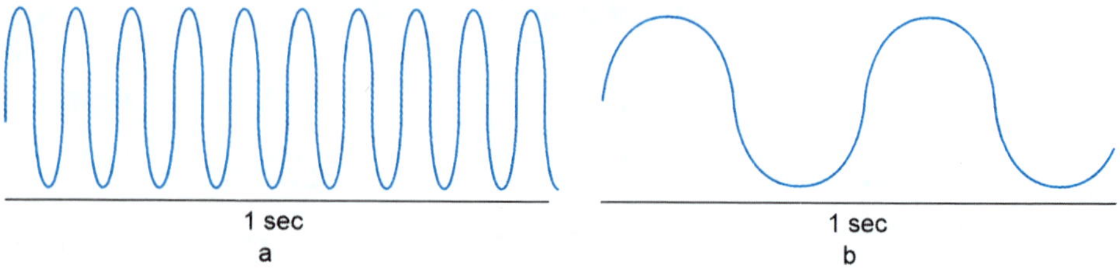

Fig. 7.1a and b: High frequency and low frequency waves

Usually, frequency is measured in the hertz unit, named in honored to 19th century German Physicist Heinrich Rudolf Hertz. The hertz measurement, abbreviated Hz, is the number of waves that passes per second. For example, an "A" note on a violin string vibrates at about 440 Hz (440 vibrations per second).

Wavelength: The wavelength of a wave is defined as the distance between two consecutive waves where the particles are travels and simply wave travels in one full wave cycle along the medium. The wavelength is frequently referred to as the length of the recurring patterns—the length of one complete wave. This length is usually measured from one wave crest to successive wave crest or from one wave trough to the next adjacent wave trough for a transverse wave. A longitudinal waves wavelength must be measured differently because it lacks crests and troughs. A longitudinal wave is made up of a sequence of compressions and rarefactions that repeats itself.

Wave speed: Wave speed is interrelation between both wavelength and wave frequency. Wavelength is the distance between successive points of waves. Wave frequency is the number of waves that pass a fixed point in a given time. The relationships between wave speed, length and frequency are shown in the following equation:

$$\text{Speed} = \text{wavelength} \times \text{wave frequency}$$

In this equation, wavelength is measured in meters and frequency is measured in hertz (Hz), or number of waves per second. Therefore, wave speed is given in meters per second, which is the SI unit for speed.

7.2.2 Point Sources

Point sources of sound is nothing but the sound much smaller than the wavelength of sound it emits, it is represented by monopole and it is also called pint source. The sound wave radiates equally in all directions. This travels equally by expanding and contracting. In acoustics, it is also called compression and rarefactions.

Inverse-Square Law: According to this law, as one moves far from a spherical wave source, the sound amplitude at the far distance gets reduce. This is due to the intensity I of the ave decreases as one moves away from the source. The intensity I is the power W by ie area A at which wave spread:

$$I = \frac{W}{A} \quad ...(7.1)$$

Assume that the sound waves are absorbed in the surrounding as it propagates away om the source. Let r is distance between source and absorber in the imaginary sphere ver which the spherical wave is spread is $A = 4\pi r^2$. Substitute in Eq. (7.1).

Then,
$$I = \frac{W}{4\pi r^2} \quad ...(7.2)$$

The above equation shows the relationship between power and intensity.

.2.3 Sound Power and Intensity

ı general, when a sound wave is propagated from the source, it transmits energy. The otal sound energy emitted by a source can be determined by source per second and it is alled sound power, W (in watts).

The sound intensity is the sound power per unit area $I = \frac{W}{A}$ (refer above)

If a source is a point source then the intensity reduces with the square of the distance according to "inverse-square law").

If a source is far then the sound pressure is proportional to the square root of intensity,

i.e.
$$i \propto p^2 \text{ or } p \propto \sqrt{i} \quad ...(7.3)$$

.3 WAVE REFLECTION

n general, when a wave is interacting with surface boundaries, it is reflected and involves change in direction of waves. For example, in the phenomena of echo, sound waves are eflected from a distant object. Later it reaches to the listener with slight delay in time. 'his is due to collision of a wave with a surface, a part of it reflected whereas the rest of it ransmitted into the second medium and it is also called transmitted wave.

The incident and refracted waves are subjected to Snell's law of refraction, whereas, the ncident and reflected waves are subject to the rules of reflection. The reflected wave is llowed to propagate along the strings (actual string motion made up of 'incident wave' nd 'reflected wave'), it reaches the right end and is reflected. The pulse exerts a force on he wall when it reaches the fixed end, and the wall, according to Newton's third law, exerts n equal and opposite force on the string. This second force causes a pulse that travels ack along the string in the opposite direction. At that time, the reflected and incident ulses have opposite signs and cancel each other out.

Endless Reflection: We call it a free end when the right end of the string is connected to ring that moves up and down on a rod without friction. When the pulse reaches the ight end of the rod, the ring advances up the rod, pulling on the string and stretching t, creating a reflected pulse with the same sign and amplitude as the original pulse. The ncident and reflected pulses reinforce each other in such a reflection, resulting in the naximum displacement at the end of the string: The ring's maximum displacement is wice the amplitude of either of the pulses.

7.3.1 Standing Waves

A standing wave, also known as a stationary wave, is a wave that oscillates in time but do not move in space due to its peak amplitude profile. Simply it is created by two waves equal amplitude and frequency that travel in opposite direction. The wave oscillation peak amplitude is constant with time at any place in space, and the oscillations at differe locations throughout the wave are in phase. Nodes are the areas where the absolute valu of the amplitude is lowest, while antinodes are the sites where the absolute value of th amplitude is highest.

7.4 MEASURES OF SOUND

Many factors influence in measure of sounds which include the sound how long it last sound frequencies or pitches and other important factor of measuring sound is intensi or volume.

If the sound intensity is measured which also refereed as sound power or sound pressu in units, called decibels. Decibels (dB) are named in honor of Alexander Graham Bell, wh is the inventor of both the telephone and the audiometer. An audiometer is a device th measures how well a person can hear certain sounds.

7.5 DECIBELS

The decibel (*dB*) is a unit which is used to express the ratio between two physic quantities, commonly levels of acoustic or electric power, or relative loudness of sound One decibel $\left(\frac{1}{10} bel\right)$ = 10 times the power ratio's common logarithm. The intensity of sound in decibels is expressed as $10 \log_{10}\left(\frac{S_1}{S_2}\right)$, where S_1 and S_2 are the intensities of th two sounds; doubling the intensity of a sound represents an increase of a little more tha 3 dB. In common usage, specifying the intensity of a sound entails a comparison of th sound's strength to that of a sound barely audible to the human ear. A 60 dB (6 bel) c sound, such as regular speech sound value.

7.6 APPLICATION

Navigation, weather forecasting, tracking aeroplanes, ships, submarines, and missiles a employed with sonar. Sound is used in a variety of ways in entertainment gadgets, suc as radios, tape players, and so on.

The real beneficiary use of sound is in communication network. Sonar and Ultrasoun Medical Imaging are the two major applications of sound waves.

Sound propagation has a wide range of essential and practical applications. Sonar, whic stands for sound navigation ranging, is one such example. Sound propagation is used b Sonar to assist ships and submarines in navigating the oceans and seas. Sonar can also b used to investigate the ocean's depths and the numerous items located under the ocear Another application is ultrasound medical imaging, which is used in medicine. It work by generating a two-dimensional black-and-white image of organs and other bodily part via sound propagation. It aids in the detection of numerous defects (e.g. tumours) as wel as the understanding of how the body works.

Echolocation: Echolocation (echo + location) is the process of sending out ultrasoun waves and interpreting the source wave and reflected wave, i.e. the sound echoes tha turn. Bats that fly at night with the help of echolocation to find flying insects to eat. They n squeak out from 9 to 200 kHz of ultrasounds every second. They can identify where e target is and how it is travelling by detecting the returning echoes. They can also steer ear of obstacles such as walls, trees, and other large objects.

ltrasound's Medical Applications: Ultrasound has a wide range of applications in edicine. This is beyond the humans audible range and human cannot hear ultrasonic aves, they can be employed at very high intensities. Ultrasound waves are utilized to fely break up kidney stones in patients using high-intensity vibrations. Ultrasound aves transfer energy, that is, used to clean medical equipment. The ultrasound scanner is ne of the most important medical applications in scanning purposes. The same scientific eory that underpins sonar is used in this gadget. It works by sending sound waves into human body and then recording the waves that are reflected back. The information from ese echoes is used to create a picture on the screen.

lusic sound: Music can be created by sound waves. Why are some sounds classified as oise while others are classified as music? Music is defined as sound with distinct pitches or ythms, i.e. periodic and regular intervals. Noise is a type of sound that has no discernible attern (uneven pattern). When musical instruments are made to vibrate at their inherent equencies, they produce pitches and rhythms. To produce varied pitches, several musical struments feature pieces that vibrate at different frequencies. The sound of the instrument created by the combination of all the pitches and the instrument's resonance.

Chapter

8

Optics

8.1 PRINCIPLES OF LIGHTING

In physics, light is a wave or a stream of particles. In Chapter 7, about waves are like soun or waves on a surface of water. Waves are formed by perturbation of periodic arrangeme of spatial pattern which travels in space. In a simple manner to imagine a wave which ca be visualized as wave by sketching the wave fronts with crests moving up and down i periodic manner (*see* Fig. 7.1).

8.2 THE NATURE OF LIGHT

Light is a transverse wave and it comes under normal visible range of electromagnet radiation that human can be seen by their normal eye (emmetropic eye). The nature light is a wave which was first demonstrated through an experiment of diffraction an interference. As we know an electromagnetic waves can travel through a vacuum similar light as do and perform polarization.

Sources: In early 16th century, light is generated by the Incandescence and Luminescen emission. In Incandescence, the emission of light is from "hot" matter which is heate greater than 800 K, whereas in luminescence, the emission of light is the excitation excited electrons that fall from higher energy levels to lower energy levels. Nowaday light is produced only by the acceleration of charged particles.

Speed: The speed of light in a vacuum is represented by the letter c from the Latin celerita Measurements of the speed of light is c, = 3×10^8 m/s.

- The speed of light is a constant in all inertial frame of reference.
- The speed of light in a vacuum is fixed at 299,792,458 m/s.
- The speed of light in a vacuum is maximum and it is always slower in a medium.
- The speed of light vary depends upon the medium. The speed of anything with ma is always less than the speed of light in a vacuum.

Amplitude: The amplitude of a light wave is associated to its intensity. Intensity is th absolute measure of a light wave's power density. Brightness is the relative intensity perceived by the average human eye.

Frequency of a light: The frequency of a light wave is associated with its color. In case monochromatic light is described by frequency of single value. For example, laser lig is effectively monochromatic.

There are few color bands are forming a band of monochromatic light that helps increase frequency they are red, orange, yellow, green, blue, and violet. Sometimes, contra visible lights are ranges from "ultraviolet light" and "infrared light". Beyond this range of electromagnetic radiation are not visible to human.

Polychromatic light is described by many different frequencies. Almost every light source is polychromatic. Even white light is also polychromatic. A continuous spectrum in electromagnetic radiation has all range of frequency. But in characteristic spectrum has discrete set of well-defined isolated frequencies. The excited electrons in a gas emit a characteristic spectrum. The wavelength of a light wave is inversely proportional to its frequency ($v \alpha c/\lambda$). Light is frequently described by the wavelength in a vacuum. Light ranges in wavelength from 4000 Å on the violet end to 7000 Å on the red end of the visible spectrum.

8.3 NEWTON'S CORPUSCULAR THEORY

The light is a form of energy that is emitted in the form of photons or form of waves. With the help of light energy human can see the object and realize the things. The nature of light to explain the phenomena such as reflection, refraction, diffraction, interference, etc. Newton's corpuscular theory was based on the following postulates:

- Newton proposed that a source of light emits many minute, elastic, rigid and massless particles and which can produce the sensation of vision.
- Depends on the source of light, they produce different colours.
- In a straight line, particles travel through a transparent medium at very high speed in all direction.
- These light particles are repelled by a reflecting surface and attracted by transparent materials.

Merits

- The rectilinear propagation of light is explained well.
- The reflection and refraction of light separately can be explained.

Demerits

- Newton's corpuscular theory fails to explain partial reflection and refraction on the surface of transparent medium.
- The corpuscular theory fails to explain optical phenomena such as interference, diffraction, polarization, etc.
- According to this theory, velocity of light is larger in the denser medium than in the rarer medium, but experimentally it is disproved
- As the particles are emitted from the source, mass of the source of light should decrease but experiment mass is always constant.

8.4 HUYGENS'S WAVE THEORY

In 16th century, **Christian Huygens** proposed his theory on mechanism of the propagation of light, nowadays known as Huygens' principle. He stated in his principle a very fundamental property of waves, which will be a useful tool to explain certain wave phenomena, like reflection and refraction of plane waves. According to his principle, all points on a wavefront act as sources of new waves, and the wrapping of these secondary waves that constitute the

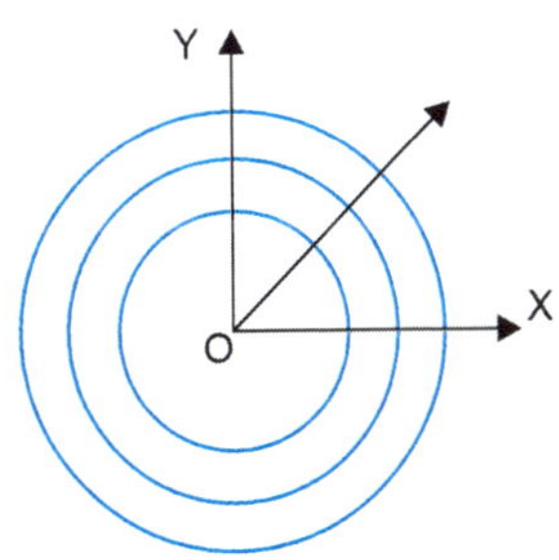

Fig. 8.1: Huygens wave theory

new wavefront. For example, when a small stone is dropped in a calm pool of water that spreads wavefront out from the point of dropped stone. Every point on the surface starts oscillating with time. At instant, if we capture the incident that show circular rings as shown in Fig. 8.1. If the impact is stronger, it produces concentric ring.

If the point source emits waves in all directions, then all the wavefronts have same amplitude and vibrate in the same phase and appears sphere shape. It is known as spherical wave and at large distance of these spherical wave is appeared as plane; it is known as plane wave. If we know the shape of the wavefront at $t = 0$, then Huygens principle helps to determine the shape of the wavefront at any time ($t = x$).

Postulates of Huygens' Wave Theory

- Light is the form of waves which is also emits only from source of light.
- Light waves are longitudinal in nature and travel like sound wave but reach faster than sound.
- In a homogeneous medium light travel with constant velocity.
- Light with different colors have differ in wavelength.
- Light waves are electromagnetic wave with different colors of light that human eye can feel the sensation of vision.

Merits

- It can explain phenomenon such as reflection, refraction, interference and diffraction.
- The wave theory of light can explain the phenomenon of partial reflection and refraction of light.
- According to theory, velocity of light is less in denser medium than the velocity of light in a rarer medium.

Demerits

- Wave theory of light considered the presence of hypothetical ether medium but experiment proved that there is no ether.
- It is failed to explain rectilinear propagation of light.
- It is failed to explain compton effect and polarization of light.
- It is failed to explain bending of wave through an obstacle.
- Wave theory of light considered that light waves are longitudinal in nature but experiment not proved.

8.5 MAXWELL'S ELECTROMAGNETIC THEORY

According to Maxwell's electromagnetic theory, light is electromagnetic waves. Experimentally, it is observed that velocity of light is equal to the velocity of electromagnetic waves and the particles are travel at the speed of light, even they travel in vacuum at the same.

Maxwell described light as a propagating wave of electric and magnetic fields. More generally, he projected the existence of electromagnetic radiation—coupled electric and magnetic fields traveling as waves at a speed equal to the speed of light.

According to this theory, light is spread in the form of wave packets with energy called quanta. Each quanta of light also called photon. The energy of photon is directly proportional to frequency of light,

i.e.

$$E \propto \nu$$

$$E = h\nu \quad \text{...(8.1)}$$

Where, ν = frequency of light and h = Plank's constant

8.6 PHOTOELECTRIC EFFECT, WAVE THEORY OF LIGHT AND WAVE-PARTICLE DUALITY

In 19th century, the wave nature of light was well established. The phenomenon of reflection, refraction, interference, diffraction, and polarization were explained with the help of wave picture of light and proved that light possess **'wave nature'**. At the same time, photoelectric effect, compton effect and the emission, absorption of radiation are the phenomena that prove light possesses a quantum or a particle nature. So, light has dual nature, which means it behaves like a wave as well particle.

Accordingly, light is an electromagnetic wave consisting of electric and magnetic fields with continuous distribution of energy.

According to the light wave picture, the free electrons at the surface of the metal can absorb the radiant energy incessantly. The electric and magnetic field of light with higher intensity of radiation along with higher amplitude. Therefore, the greater the intensity of radiation can absorb higher energy by the electron. In this context, the electron with maximum kinetic energy is then projected to rise with increase in intensity. In addition, over the entire wavefront of the radiation the absorption of energy by electron takes place incessantly. Since many electrons absorb energy, energy absorbed by electron per unit time will be small. So that it can take hours or more for a single electron to pick up sufficient energy to overcome the work function. The photoelectric emission is an instantaneous process without any apparent time lag. The wave picture is failed to explain the most basic features of photoelectric emission.

Energy Quantum of Radiation—Einstein's Photoelectric Equation: In 1905, Albert Einstein (1879–1955) projected a new picture of electromagnetic radiation to explain photoelectric effect. In this wave picture of light, photoelectric emission does not take place by continuous absorption instead it is discrete units. So, it is called quanta of energy of radiation. A quantum of particles has an energy $h\upsilon$, where υ is the frequency of light and h is plank's constant. In photoelectric effect, an electron of radiation absorbs a quantum of energy ($h\upsilon$). If it is absorbed minimum energy even that is more sufficient to eject electron from metal surface. The ejected electron has maximum kinetic energy.

$$K_{max} = h\upsilon - \phi_0 \qquad \text{...(8.2)}$$

The electron is tightly bounded in the metal surface, after ejection it has kinetic energy which is less than the maximum value ($h\upsilon - \phi_0$). But the intensity of light is determined only by the number of impact photons. Intensity of light is also depending on number of electrons ejected per second. Each individual incident photon energy also plays a major role in the maximum kinetic energy of the ejected photoelectrons. The above equation (8.2) is known as Einstein's photoelectric equation.

8.7 PROPERTIES OF LIGHT

In physics, light is the form of energy in which objects are made to visible for human eye. Light is an electromagnetic radiation that can be perceived by human. The important properties of light are as follows.

i. Diffraction,
ii. Dispersion,
iii. Interference,
iv. Polarization,
v. Reflection, and
vi. Refraction.

In addition, there are some important aspects of light.

- Light is a form of energy.
- It is always travels along a straight line.

- It does not require any medium to travel. Even it can travel through vacuum.
- It can propagate very faster, the speed of light in air is 3 lakhs km per second.
- Light colors can be differentiate by its wavelength and frequency.
- In white light dispersion, violet has lowest wavelength and red has highest.

8.8 PHOTOELECTRIC EFFECT

The photoelectric effect is the process of generating a photoelectron by ejecting electrons from the surface of a metal. This happens only when light or electromagnetic radiations (such as X-rays, ultraviolet rays) is incident on the metal surface. *This process of release of electrons from a metal with the help of light photon at suitable wavelength, is called the photoelectric effect.*

These ejected electrons are called photoelectrons. The frequency of the incident light has a vital role in photoelectric emission. The emitted photoelectron's kinetic energy depends on the incident photon light frequency. The process of emission of photoelectrons from metal surface are commonly referred as photoemission.

Schematic diagram of photoelectric effect and emission of photoelectron is shown in Fig. 8.2.

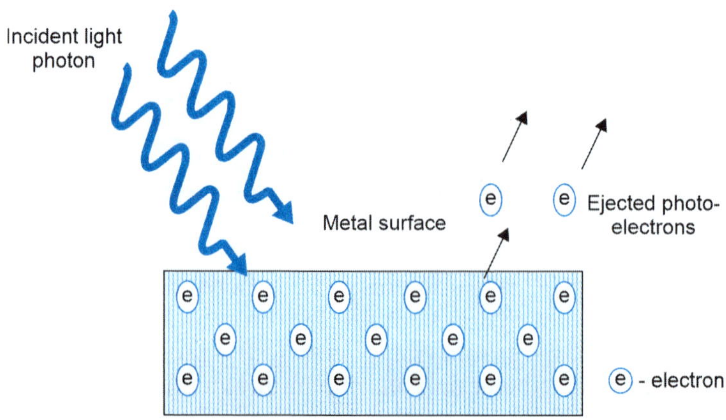

Fig. 8.2: Schematic diagram of photoelectric effect

Detail Description of Photoelectric Effect—Photon Concept: Let us consider a photon, the wave nature of light. The concept of photoelectric effect can be explained through the photon of light. Light can be imagined a group of particles called photons. The energy of a photon is depended on the frequency of the light and given by the following Planck's equation:

$$E = h\nu = h\left(\frac{c}{\lambda}\right) \quad ...(8.3)$$

Where,

E denotes the energy of the photon

h is Planck's constant

$\nu = \frac{c}{\lambda}$ – frequency in inversely proportion to wavelength

c is the speed of light (in a vacuum)

λ is the wavelength of the light

From Eq. (8.3) different frequencies of light have photons of different energies and wavelength. For example, the frequency of violet light is greater than that of red light (the wavelength of violet light is shorter than the wavelength of red light). So that, the energy of photon of blue light is greater than the energy of a photon of red light.

8.9 REFLECTION, REFRACTION, AND DISPERSION

Reflection: Let us consider a ray of light that comes back from the medium, where the light is striking on the shiny surface. The returned light towards the origin or some other direct or not 180° to incident light then the phenomenon is called reflection of light. Simply it is the bouncing back of the light rays from a surface in the same medium.

8.9.1 The Law of Reflection

It states that the angle of incidence and angle of reflection are equal. The incident ray, reflection ray and normal is at the same plane (Fig. 8.3).

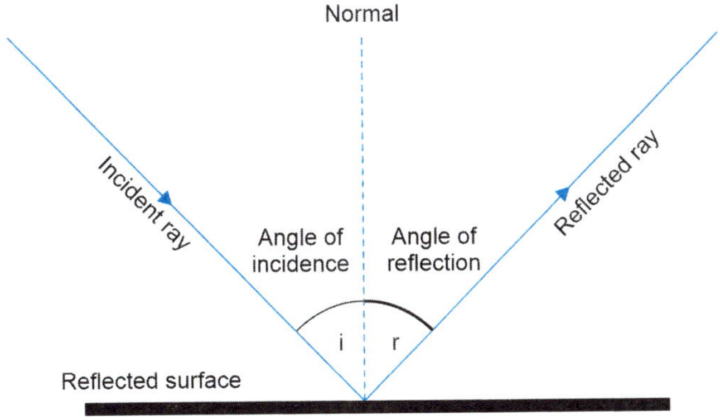

Fig. 8.3: Law of reflection

- Reflection is the returning back of light when it hits on a smooth surface.
- The nature of reflection is generally happening only on shinny surfaces that allow light recovering without penetration.
- Reflection has two forms, that are regular reflection (specular reflection) and diffused reflection.
- Reflection light returns in the same direction.
- During reflection, the speed of light is independent, i.e. speed does not vary
- During reflection, the angle of incidence and angel of reflection is same ($i=r$)

Regular reflection: If the beam of light rays falls on a polished, regular, and smooth surface, regular reflection occurs. It is also termed as specular reflection.

Irregular reflection: If the beam of light rays falls on an unpolished, irregular, and rough surface, irregular reflection occurs and light reflected in various directions. It is also termed as diffuse reflection.

Refraction: Let us consider a ray of light, that incident on a medium where the direction of light changes and the speed of light vary as it enters from one medium to another (e.g. air to transparent medium). It is called refraction of light. Examples of refractions are lenses, glasses, prisms, telescope, etc.

8.9.2 The Law of Refraction

- It states that "the ratio between the sine of angle of incidence and the sine of angle of refraction is constant at any instance of two given medium".

- The incident ray, normal and the refracted ray are at the interface of any two given medium (Fig. 8.4).

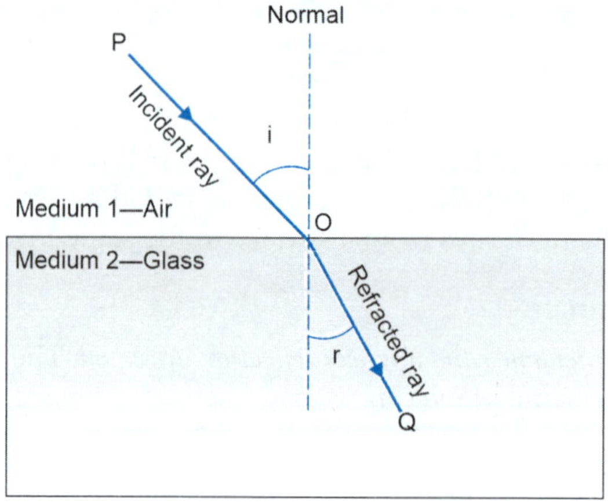

Fig. 8.4: Law of refraction

- Refraction is the deviation or bending of light from the path when it travels from one medium (air) to another (glass).
- The nature of refraction is generally happening only in transparent surfaces. Bending of light happens when the ray travel from one medium to another.
- The speed of light is dependent on the medium that travel, i.e. velocity of light will vary depends on the medium, for example; in air, $c = 3 \times 10^8$ m/s whereas in the denser medium, c is low.
- During refraction, the ratio between sin of angle of incidence and sin of angel of refraction is constant.
- Refraction index $= \dfrac{\sin i}{\sin r}$

The refractive index is the ratio of the angle of incidence to the angle of refraction. It is also defined as the ratio between speed of light in vacuum (air) to medium. It determines the speed of a ray of light in the new medium. This implies that denser the medium the slow is the speed of light, and vice versa.

8.9.3 Dispersion

The scattering of white light into different forms of colors, i.e. full spectrum of wavelengths is called dispersion. It occurs only when the direction of light is changed and it depends on wavelength of light.

In a simple way, the dispersion is nothing but the deviation of light from the path and the point of incident. If the light disperse into different colors depends on the wavelength. For a given medium, the spectrum of light where deviation of light increases as wavelength (λ) decreases and λ is greatest for violet light. Thus, violet light is bent more than red light, as can be seen with a prism (Fig. 8.5).

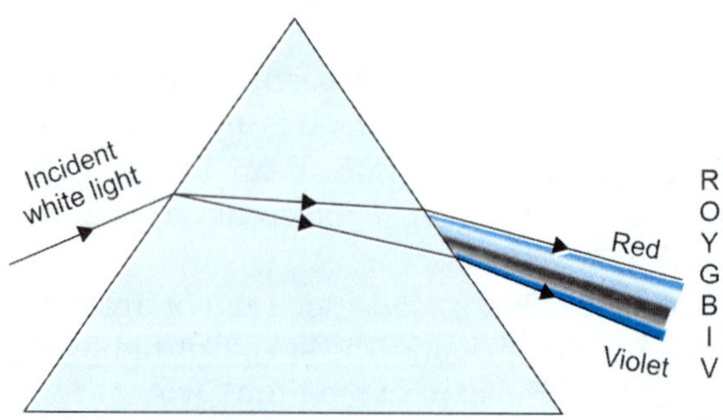

Fig. 8.5: Dispersion of white light in glass prism

Similarly, in case of a rainbow, light enters a drop of water and is reflected from the back of the drop. The light is refracted both as it enters and as it leaves the drop.

White light is dispersed by the prism. Since the refractive index is varies with wavelength of light so the refractive angles also vary with wavelength. A sequence of red to violet is produced, because the index of refraction increases steadily with decreasing wavelength.

8.10 TOTAL INTERNAL REFLECTION AND FIBER OPTICS

Total internal reflection is the process that occurs only when the light rays are travel from optically denser medium to optically lighter medium. During the process of internal reflection, the light rays are partly reflected back into the same medium, whereas the remaining par in refracted to the second medium. This is called internal reflection.

Total internal reflection should satisfy the following

Light ray travels from denser to rarer medium

- If the light rays are moving from a denser medium to a rarer medium, it moves away from the normal. The ray will get refracted at an r (angle of refraction).
- In this case, increasing angle of incidence from normal that causes the angle of refraction become less.
- If the angle of incidence in keep on increasing and at an angle of 90°, it will become equal to the angle of refraction. Beyond that angle there will be no refraction.

Limitation

- The angle of incidence (i) should be greater than the angle of incidence for which angle of refraction is 90°.
- Angle of incidence should correspond to angle of refraction = 90°. This angle is known as critical angle.
- Angle of incidence should be greater than critical angle. The critical angle is only defined when the ratio between refractive indices is less than 1.

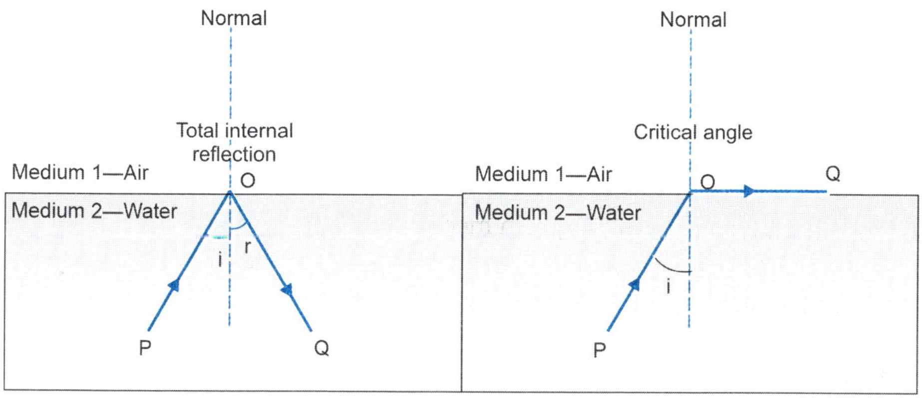

Fig. 8.6: Total internal reflection

8.11 OPTICAL FIBER

One of the major application of total internal reflection is an optical fiber. It can be used to restrain light. It is a thin, transparent fiber, that is usually made of glass or plastic, for transmitting light impulse. The construction of a single optical fiber is shown in Fig. 8.7.

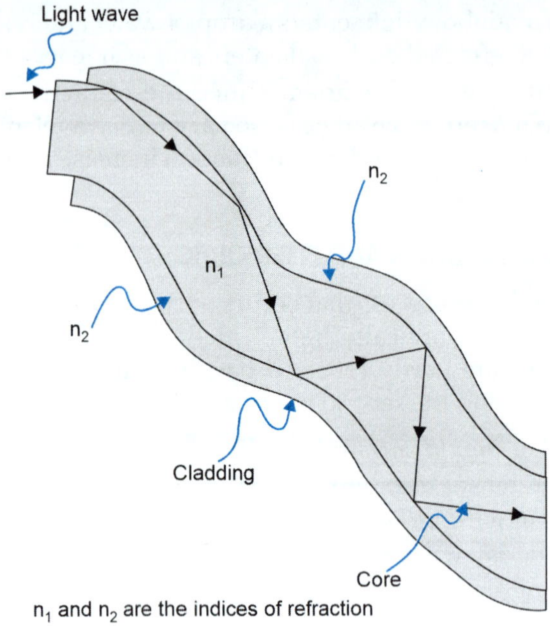

Fig. 8.7: Total internal reflection in an optical fibre

The optical fiber consists of outer and inner layer which is a protective cladding and light reflector called core respectively. Clad has lower index of refraction and core can ensure the total internal reflection. The total diameter of the fiber is around 125 μm and the core alone is about 50 μm. The difference in cladding and core refractive index can allows total internal reflection like air-water interface medium. When the angle of incidence is greater than the critical angle, the internal reflection occurs. So that the light is not allowed to escape outside and it is trapped inside the fiber only. In this way light travels very quickly to over a length of fiber. As light can travel with a velocity of 3 lakhs km per sec, so it can be travelled for very long distance.

Applications

- The most common application of optical fibers is used in telecommunications, because it can transported the information over long distances with a short period of time and with minimal loss of data.
- Another common use can be found in medicine like endoscopes.
- The field of applied science and engineering also concerned with the design and application of optical fibers which is also called fiber optics.

8.12 LASER

Laser is donated by **l**ight **a**mplification by **s**timulated **e**mission of **r**adiation. It is a device that can emit light (electromagnetic radiation) at specific wavelengths through a process called stimulated emission and producing narrow beam of radiation. It has some notable properties.

- Laser light is monochromatic
- Laser light is coherent source with all waves are in phase with one another

- Laser beam is converged coherent beam and hardly diverges
- Laser beam is more intense when compared with other source.

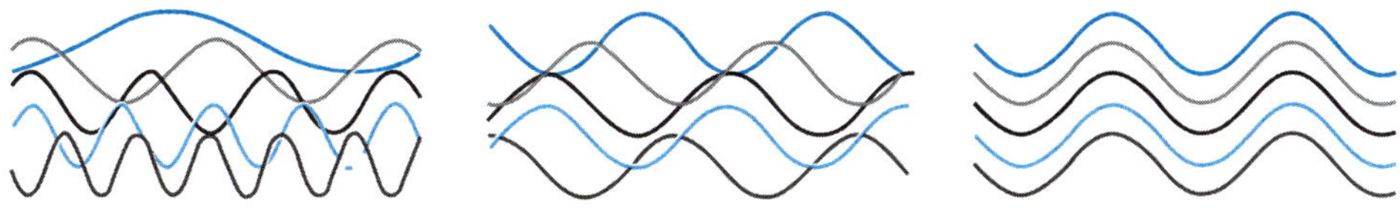

Fig. 8.8: Monochromatic, incoherent and coherent light sources

Principles of a laser: In laser photons can interact with atom in three different ways,

1. Absorption of radiation
2. Spontaneous emission
3. Stimulated emission

Absorption of radiation: It is process in which the electron can absorbs energy from photon, electron in the ground state jumps into the higher energy state. The electrons present in the orbit which is near to the nucleus, called lower energy state electrons. Usually, the lower energy state electrons require energy to jump to next level which is obtained by various energy sources such as heat, electric field or light, etc.

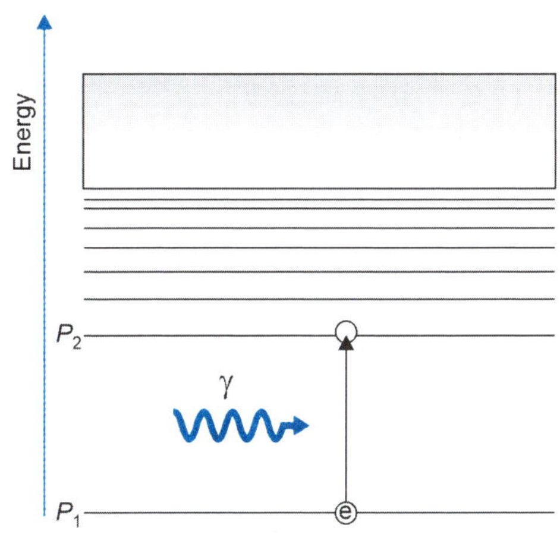

Fig. 8.9: Absorption

Let us consider two energy levels of electrons are P_1 and P_2. P_1 is the ground state energy of electrons and P_2 is the excited state energy of electrons. Usually, electrons cannot jump into the higher energy level. It requires energy to jump into the next energy level. That energy can be determined by the difference in two levels (P_2–P_1). This energy is more sufficient to remove the electron from the lower energy level to next energy level.

Spontaneous emission: It is the process in which the electrons lose it energy from excited state and return to the ground state by emitting photons. Usually, the electrons in the excited state cannot stay for a long time. In a short period of time, it will return back to low energy state. The time period of electron in the excited state is 10^{-8} second (Fig. 8.10).

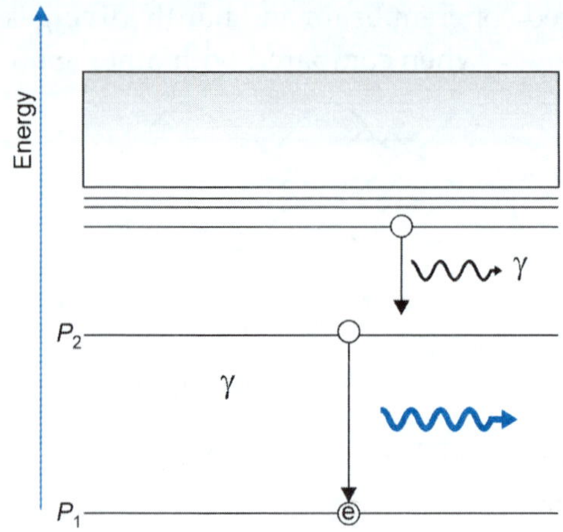

Fig. 8.10: Spontaneous emission

In spontaneous emission, when the electron loses it energy and falls to lower energy state with release of energy in the form of light. The photons which are emitted in spontaneous process create incoherent light. These incoherent lights are in random phase between them.

Stimulated emission: It is process in which incident photon interact with excited electron to force it back to ground state. In stimulated emission, the light energy is given directly to the excited electron by an artificial process. In this process, excited electron need not stay for completion of lifetime. Artificially forced to shift in ground state. In addition, the forced electron to the ground state produces an energy that again process to produce one more excited electron. This technique is called stimulated emission (Fig. 8.11).

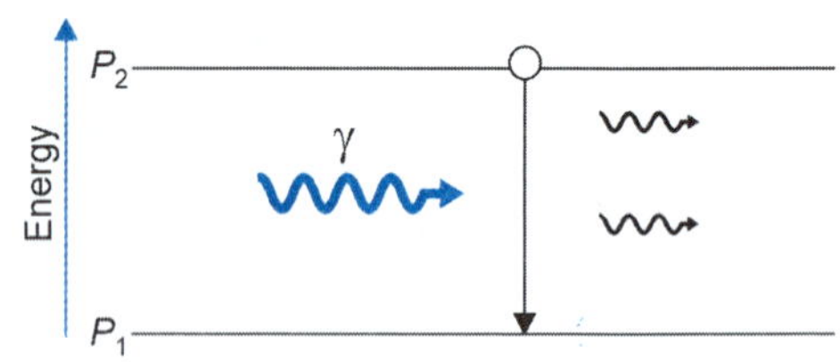

Fig. 8.11: Stimulated emission

In stimulated emission, two photons are emitted. One is due to incident photon and another in due to energy released by excited electron which causes to emit another one electron. Thus, two photons are emitted. The stimulated emission process is rapid when compared to the spontaneous emission process.

All photons which are emitted by stimulated emission have the same energy, same frequency and all are in phase. Therefore, all photons in the stimulated emission travel in the same direction. Photons which are emitted by stimulated emission is mainly depends on the number of electrons in the excited state and the intensity of incident light.

8.13 SCATTERING OF LIGHT

The scattering of light occurs when the light travels through atmosphere where different gases present. The light is refracted through all possible directions and it is called scattering of light. If the light is travel through a medium which consists of particle are called as 'scatterer'.

8.14 TYPES OF SCATTERING

Based on the energy of light beam, scattering can be classified as follows: (i) Elastic scattering and (ii) inelastic scattering.

- **Elastic Scattering:** The energy of incident beam and scattered beam of light are same, then it is called elastic scattering.
- **Inelastic Scattering:** In this case, the energy of incident beam and scattered beam of light are not same, then it is called inelastic scattering. It can be differentiated based on nature and size of scattering.
- **Mie Scattering:** When the incident light wavelength is equal or larger than the scattered light is called Mie scattering. It is also called elastic scattering. The wavelength of scattering is independent to the amount of scattered radiation. This type of scattering can be seen in dust, smoke and very low particles in the atmosphere. For example, when white light is dispersed or scattered to form different colors and all these colors combined together forms white light. This happened in the process of Mie scattering. Even white color appearance of clouds is also Mie scattering.
- **Raman Scattering:** A beam of single wavelength of light which contains only one color that passes through a gas or liquid or transparent medium, where a part of light ray is scattered. But the scattered light has some additional frequencies (or wavelengths) which is differ from incident frequency (or wavelength). This is called Raman scattering or Raman effect.

  "The monochromatic light is interacted with particles of liquid or gas that causes a change in wavelength or frequency of light that scattered" is called Raman scattering. In simple words, the scattering of light in a particle of medium without change it wavelength is called 'Rayleigh line', whereas the frequency of spectral lines differ from the incident frequency are called 'Raman lines'. If the lines of frequencies lower than that of the incident frequency are called **Stokes lines** and the lines of frequencies higher than that of the incident frequency are called **anti Stokes lines**.
- **Rayleigh Scattering:** It is defined as the scattering of light through atom or molecules of gases is known as Rayleigh scattering. For example, the scattering of sunlight by the atoms of gases in the atmosphere.
- **Rayleigh's Scattering law:** It states that, *the amount of scattering of light is inversely proportional to the fourth power of its wavelength*

  i.e. $$'S' \propto \frac{1}{\lambda^4}$$

  According to this law, if the wavelength is less, then the scattering of light is increase in the 4th power. For example, when sunlight of shorter wavelength colors is scattered much more than the longer (red) wavelength colors. This is reason why the sky appear in blue in color.

  During at sunrise or sunset, the white light from the sun need to travel for long distance in the atmosphere than at noon. In white light, blue colors are scattered before it reaches the earth, when the sun is reaches to earth only red light is reached. That is why the color of the sun is red at sunrise and sunset.
- **Tyndall Scattering:** If a beam of light is entered into the smoked or dusty room through a hole or window, then the path of light is visible. This is due to the tiny particles in the dust can scatter the beam of light. So the path is visible, similarly, if the light is pass through dust solution where we can observe the path of light. This effect is called Tyndall scattering or Tyndall effect.

Blue color appearance of eye is also a Tyndall effect. Usually, the color of eye is mainly depending on the presence of amount of melanin pigment in iris. In this case of blue color eye, melanin is relatively lower than the black, brown color eyes. When light is incident on iris, it is scattered due to the Tyndall effect.

8.15 LENSES

Lens is an optically transparent medium which has two spherical refracting surfaces—one plane and one spherical surface. It is classified into two types, they are: (i) Convex lens and (ii) concave lens.

8.15.1 Convex or Bi-convex Lens

Lens under this type has relatively thick at the center and thin towards corner and upper edges. The edges are also curved outward. When a beam of light passes through this lens, it converged it into a point which is in principle axis. So, it is also called converging lens or convex lens.

8.15.2 Concave or Bi-concave Lens

Lens under this type has two spherical surfaces in which lens is thinner at the centre than at the edges. When a parallel beam of light passing through this lens, it diverged or spread out parallel light rays. So, it is called concave lens or diverging lens.

Other types of Lenses

- **Plano-convex Lens:** In convex lens, one-side of bi-convex lens is plane and other side is spherical curved shape. This type of lens is known as a planoconvex lens.
- **Plano-concave Lens:** In concave lens, one-side of bi-concave lens is plane and other side is spherical curved inward shape. This type of lens is known as a planoconcave lens. All types of lenses are shown in Fig. 8.12.

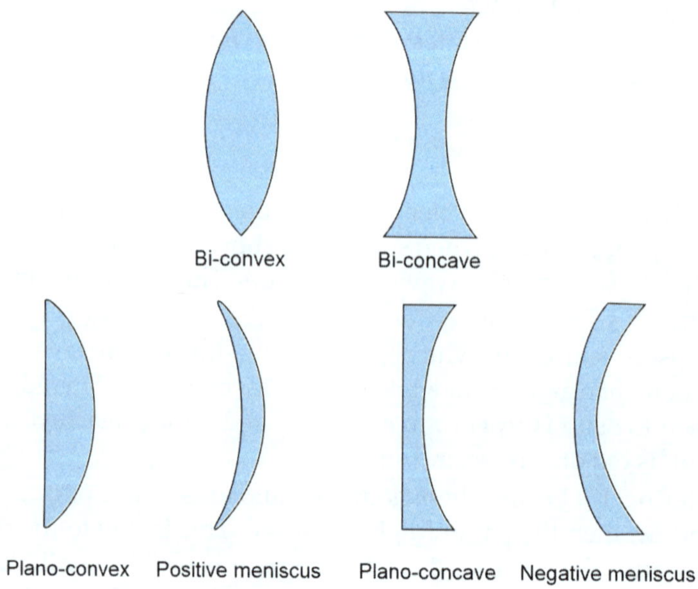

Fig. 8.12: Types of lenses

8.16 REFRACTION THROUGH CONVEX AND CONCAVE LENSES

Let us consider an object that placed in front of a convex lens, the object in the form of light rays fall on the lens. The image formation, position and nature of image can be understood by knowing the following basic rules.

Rule (i): When the light rays are passes through the convex or concave lens obliquely at its center axis (optical centre), it will continue to its path without any deviation (Fig. 8.13a and b).

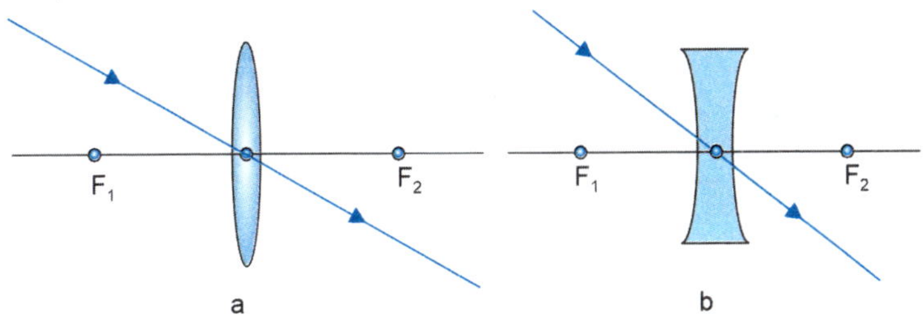

Fig. 8.13a and b: A ray of light passing through the optical center

Rule (ii): When the light rays that are parallel to the principal axis passes through the convex or concave lens, the rays are refracted from convex lens and are converged or diverged from concave lens. In convex lens, it converges to point, called focal point F_2 and in concave lens, it diverges away (Fig. 8.14a and b).

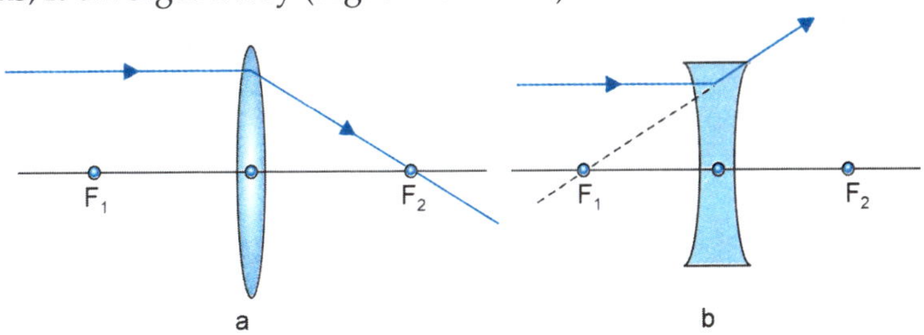

Fig. 8.14a and b: A ray of light passing parallel to the optical axis

Rule (iii): When a ray is passing through principal focus F_1 of convex lens or directed towards F_2 principal axis in concave, after the refraction, the light ray will be parallel to the principal axis (Fig. 8.15a and b).

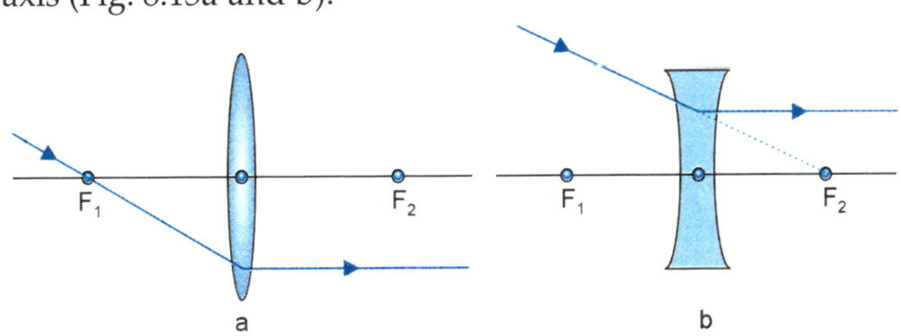

Fig. 8.15a and b: A ray of light passing through the principal focus

8.17 REFRACTION THROUGH A CONVEX LENS—AN IMAGE FORMATION

Convex lens is a convergence lens that can create image when the object is placed at various positions.

i. **Object positioned at infinite distance:** Let us consider a lens 'PQ' where an object that placed at infinite distance which means very far distance (*in simple, object is in far distance as form of parallel light rays*), a real image is formed by convex lens at the focal point F_2 as shown in Fig. 8.16.

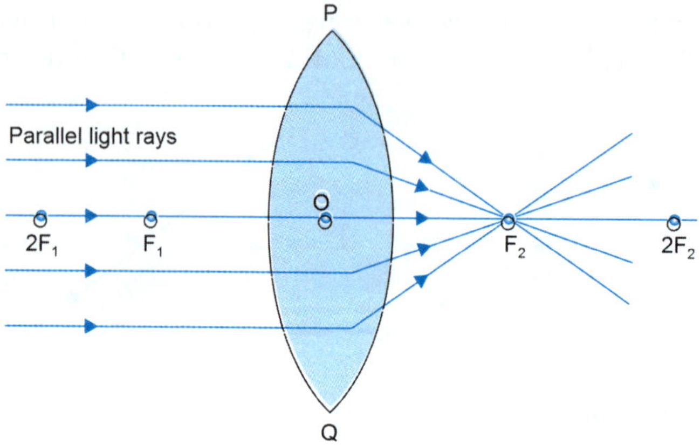

Fig. 8.16: Object positioned at infinite distance

ii. **Object positioned beyond $2F_1$:** Let us consider a lens 'PQ' where an object 'AB' is positioned beyond $2F_1$, a real and inverted image is formed in between F_2 and $2F_2$ principal focus. The image is formed as a real but inverted image 'A^1B^1' as shown in Fig. 8.17. *f* is the focal length of lens in meter.

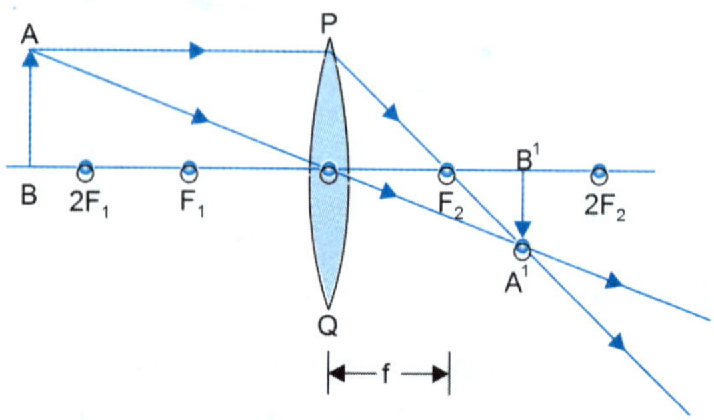

Fig. 8.17: Object positioned beyond $2F_1$

iii. **Object positioned at $2F_1$:** Let us consider a lens 'PQ' where an object 'AB' is positioned at $2F_1$, a real and inverted image is formed in exactly at $2F_2$ of principal focus. The image is formed as a real but inverted image 'A^1B^1' as shown in Fig. 8.18.

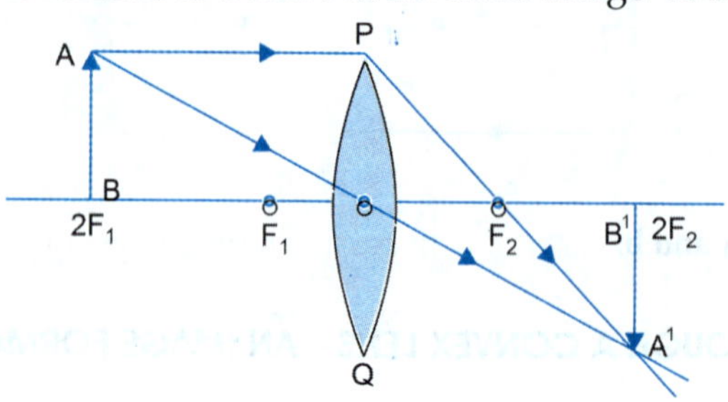

Fig. 8.18: Object positioned at $2F_1$

iv. Object positioned in between F_1 and $2F_1$: Let us consider a lens 'PQ' where an object 'AB' is positioned in between F_1 and $2F_1$, a real and inverted image is formed behind $2F_2$ of principal focus. The image is formed as a real but inverted image 'A^1B^1'. The size of the image is larger than that of the object as shown in Fig. 8.19.

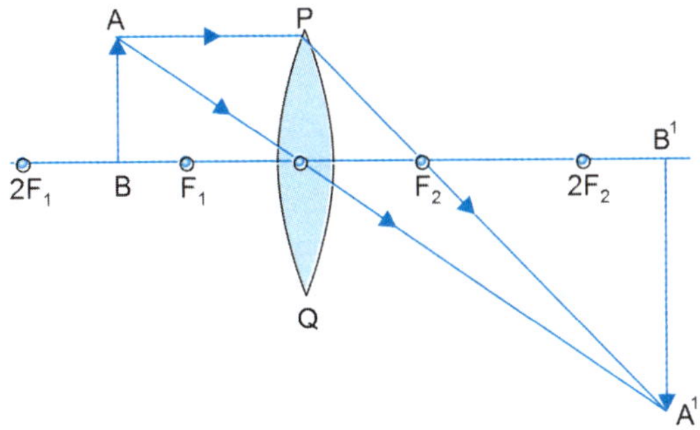

Fig. 8.19: Object is in between $2F_1$ and F_1 on the principal axis

v. Object positioned at F_1: Let us consider a lens 'PQ' where an object 'AB' is positioned at F_1, a real inverted image is formed at infinite position. The size of image very large when compared with object as shown in Fig. 8.20.

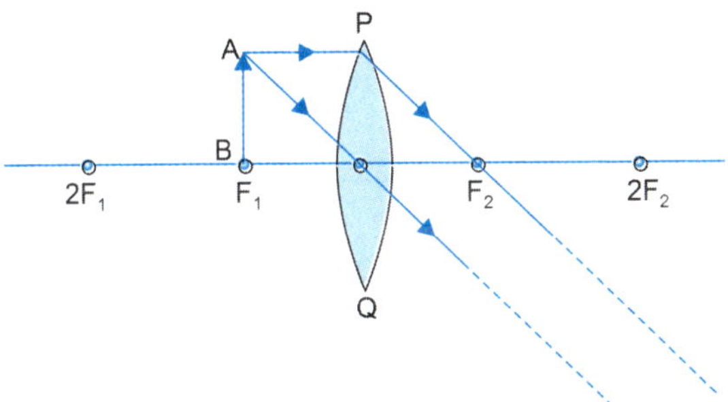

Fig. 8.20: Object is at a principal focus F_1

vi. Object positioned in between F_1 and optical center 'O': Let us consider an object 'AB' is positioned in between F_1 and optical centre O of lens, a virtual and erected image is formed. The size of image is larger when compared with object as shown in Fig. 8.21.

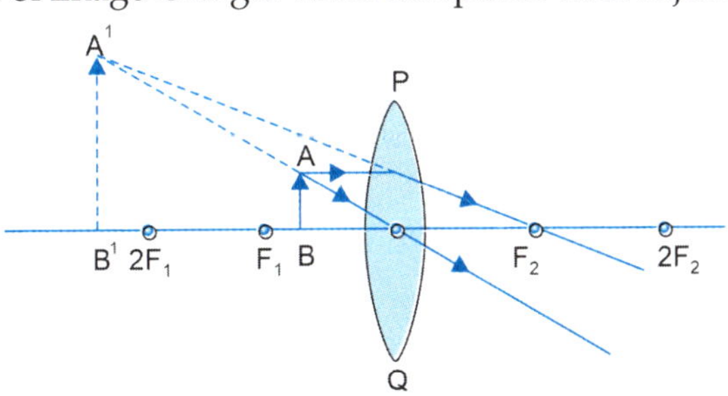

Fig. 8.21: Object is in between F_1 and optical center

Uses of Convex Lenses

1. Camera lenses are convex lenses which is major application in photographic industry.
2. It is used as magnifying lenses.
3. One of the major applications is used in microscope, telescope, and projectors.
4. Refractive errors can be rectified, e.g. correction of hypermetropia.

Refraction through concave lens: Let us consider a concave lens 'PQ' where an object that placed in concave lens, the image formation of lens in done by placing the object at two different possible positions that are as follows.

i. **Object positioned at infinite distance:** Let us consider an object that placed at infinite distance, a virtual image is formed at F_1 as shown in Fig. 8.22. The size of the image is much smaller than that of the object.

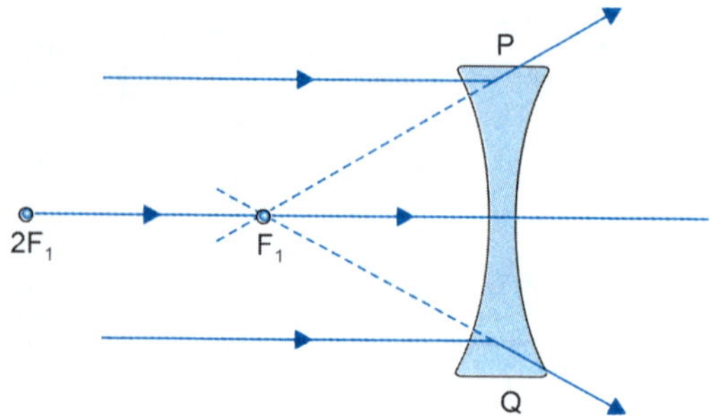

Fig. 8.22: Object is at infinite distance

ii. **Object positioned in between F_1 and $2F_1$:** Let us consider an object 'AB' is positioned in between F_1 and $2F_1$, a virtual image is formed between optical center and F_1 of the concave lens. The size of the image is smaller than that of the object as shown in Fig. 8.23.

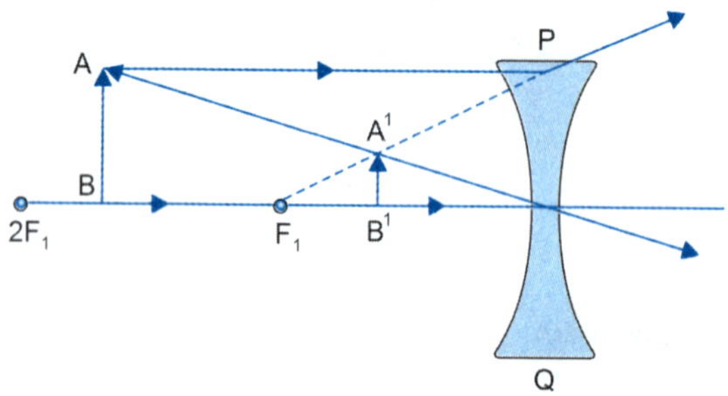

Fig. 8.23: Object is in between F_1 and $2F_1$

Uses of Concave Lenses

1. It is used in glasses for the correction of refractive error myopia
2. It is used in 'Galilean telescope'
3. It is also have an application in cameras, lasers, flashlights, peepholes, spy hole in doors, etc.

8.18 LENS FORMULA

In optical science, the lens formula gives the relationship between the distance of an object (u), distance of the image (v) and the focal length (f) of the lens in meter. It is stated as follows:

$$\frac{1}{f} = \frac{1}{v} - \frac{1}{u} \quad ...(8.4)$$

The lens formula is suitable for both convex as well as concave lenses. The important point to note that is the lenses should have negligible thickness. Equation (8.4) is known as the *lens formula*.

8.19 CARTESIAN SIGN CONVENTION OF CONCAVE AND CONVEX LENS

According to Cartesian sign convention, a convex lens focal length is always positive value, whereas a concave lens focal length is in negative value. During demonstration, the incident light always moves from left to right because the object is always placed on left side of lens. If the distance measured from the optical centre of the lens to the right side is positive. The refraction of light in concave and convex lens has the following sign convention:

- All distances are measured from the optical centre of the lens.
- Distances measured in the direction opposite to the direction of incident light is considered negative.
- Distances measured in the same direction as that of incident light is considered positive.
- Heights measured upwards and perpendicular to the principal axis is considered positive.
- Heights measured downwards and perpendicular to the principal axis is considered negative.

8.20 MAGNIFICATION OF A LENS

Magnification refers to the increase in image size created by spherical mirrors in comparison to the size of the object. It is the height of the image divided by the height of the object, and it is denoted as '*m*'. A spherical mirror's magnification, *m*, can be represented as follows:

$$m = \frac{h}{h'}$$

Where, h is the height of image and h' is the height of the object

8.21 POWER OF A LENS

The power of lens is mainly depending on the ability of lens to converge or diverge the light rays. So, the ability of lens and it degree to converge or diverge light ray that define the power. In numerically, it reciprocal of its focal length.

$$\text{Power} \qquad P = \frac{1}{f}$$

Where, f is the focal length of lens in meter.

The power of the lens if measured in diopters (D). Diopter is the unit of refractive power of a lens. For example, if an object is placed in the left side of lens and it is focused at a focal point F_2 which is at 1 meter distance as shown in Fig. 8.24b. Then the power of the lens is 1 diopter $\left(\frac{1}{1}\text{ meter}\right)$ or 1 m^{-1}. Similarly, if lens is focused an object at distance 0.5 meter of focal point F_2 in Fig. 8. 24a. Then the power of the lens is 2 diopters $\left(\frac{1}{0.5}\text{ meter}\right)$ or 2 m^{-1}. If focal length is decreased only when the power of the lens is increased.

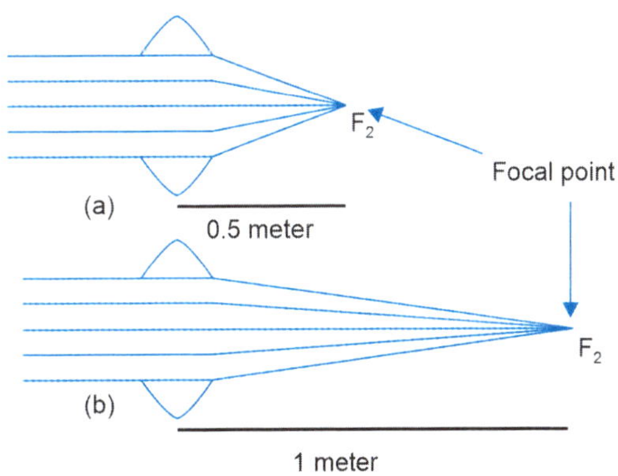

Fig. 8.24a and b: Power of a lens

8.22 LENS ABERRATION

In optical physics, aberration is one of the major properties of lens in which light rays are spread out instead of focus at a point. This is called aberration and it causes on an image that not sharp and forms unclear image. In general, aberration may cause many reasons such as lens appropriation, thickness, size, material, and positions, etc. In simple words, lens is failed to focus of light from any given point of object in a single point. Poor performance of light due to inadequate proper lens property in image focus is called aberration.

8.23 TYPES OF ABERRATION

There are some aberration taken into the consideration for the types of monochromatic aberrations.

- Astigmatism
- Coma
- Curvature of field and distortion
- Spherical and chromatic aberration

8.23.1 Astigmatism

In optical physics, astigmatism is a failure of lens to focus the image at a single point and in case of biophysical way of explanation of astigmatism is a type of refraction errors. This error is mainly due to imperfection of the meridian plane and lens in not having proper curvature. So the object is focused in one side of curvature that created one focal point, the same object is focused in perpendicular meridian, i.e. other side of curvature that created another focal point as shown in Fig. 8.25. In astigmatism, there are two focal point, for example, if a person with astigmatism seeing a object like star, that is, not visible like normal and it will be observed like a radiation star by astigmatism eye. This is due to improper curvature of human eye lens which is failed to focus an object on the retina. So, the normal star is appeared for them like radiating star.

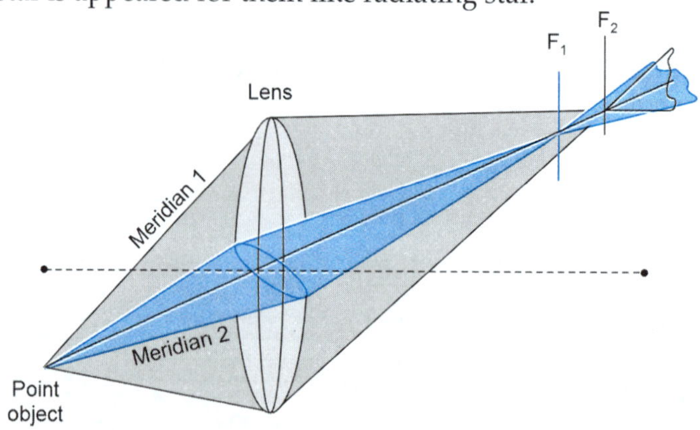

Fig. 8.25: Astigmatism in lens

8.23.2 Coma

The term "coma" is termed for the lens aberration due to imperfection of lens to focus a point image and it is blurred into the appearance of comet shape as shown in Fig. 8.26. When rays are focused by lens that produce comet image due to the variance in magnification of lens. There are two types of coma—positive and negative.

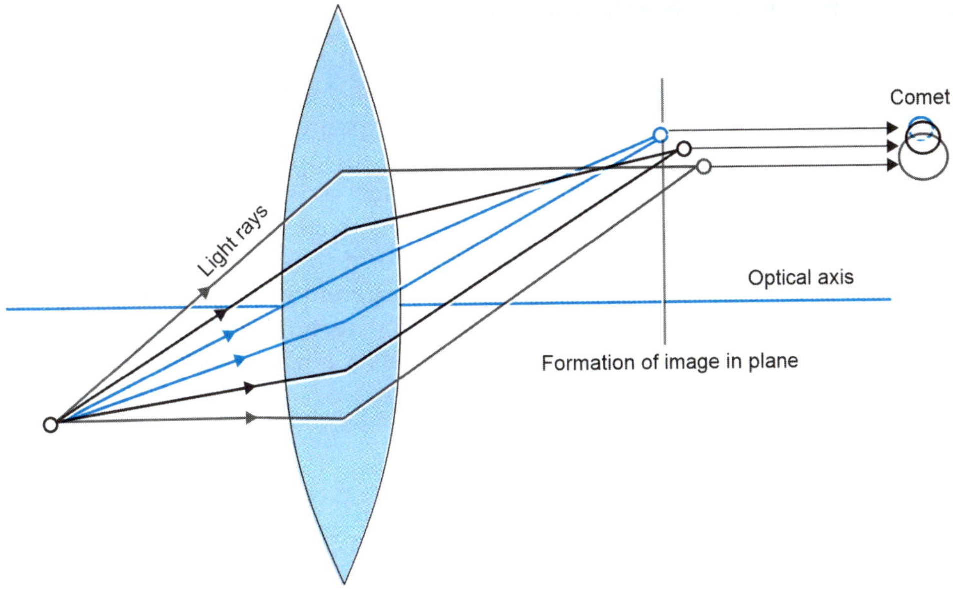

Fig. 8.26: Lens aberration coma

The tail of the comet shape is mainly depended on the type of comatic aberration. If the shape lens is imperfect, then meridional rays passing through its periphery and the image plane is closer the axis when compared to the rays which is near to axis.

If the rays produce the smallest image, it is termed as negative comatic aberration. In case of peripheral rays are focused further down the axis that causes greater magnification, it is termed as positive comatic aberration. In negative coma, the tail is pointing away from the field of view, whereas in positive coma, it points towards the field of view. A strapping concave positive meniscus lens will produce negative comatic aberration, whereas a bi-convex or plano-convex lens will produce negative coma.

8.23.3 Curvature of Field and Distortion

In aberration of lenses, curvature of field is occurring when the sharp image is focused and formed on a curved surface instead of flat plane. It also called field curvature or Petzval field curvature. Usually, objects are focused in normal surface, where it creates sharp picture, whereas in this type, optical aberration happens due to the curved nature of an optical elements. Distortion, on the other hand, refers to the deformation of an image. Distortion is an aberration commonly seen in stereoscopic microscopy, and is manifested by changes in the shape of an image rather than the sharpness or color spectrum. There are two kinds of deformation, namely barrel distortion and pincushion distortion.

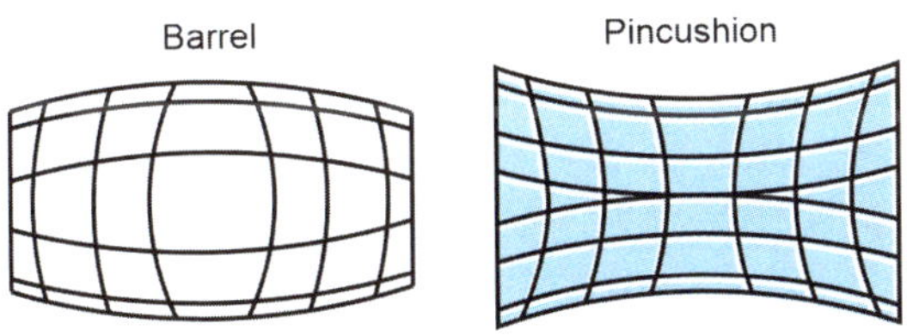

Fig. 8.27: Lens distortion

8.23.4 Spherical and Chromatic Aberration

If lenses are made with transparent material of spherical surfaces, the light rays which are parallel to the optic axis of lens focused on the axis but at different distances from the optic axis and it fails to converge to the same point. Due to the variation in the surfaces of lens, the focal point is in the range of the optical axis where the images are formed. Instead of single focal point in optical axis, its formal series of point like a small area in the optical axis as shown in Fig. 8.28. For a thick lens made from spherical surfaces, the focal distance will differ for different rays, and this change is called spherical aberration. The focal length for different wavelengths will also differ slightly, and this is called chromatic aberration. Correction for this aberration can be done by arranging the central part of the lens has a higher refractive power which helps in focusing all light rays at one point.

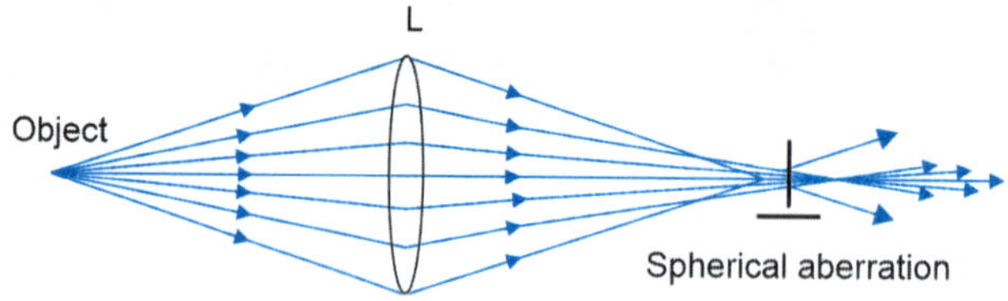

Fig. 8.28: Spherical aberration

Chapter

9

Medical Physics

9.1 BLOOD PRESSURE MEASUREMENT

The measurement of blood pressure is one of the major diagnosis and checks that your doctor performs routinely in treatment part. Problems in cardiovascular system are frequently reflected by abnormal blood pressure readings. The measurement of blood pressure is the specific direct diagnostic procedure in identifying malfunction of the heart. During operations BP is continuously monitored in intensive care units in the hospital. The integrity of the heart valves can be identified by the measurements of blood pressure

The subject at normal resting state will have a heart rate of 70–80 beats per minute approximately. The amount of blood is pumped out by left ventricle of heart ejects about 70 ml of blood into the aorta per beat during systole and also called stroke volume. At the interruption of each stroke the pressure within the arteries reaches a maximum and then declines during the rest of the cardiac cycle that means during systole (contraction of heart) pressure rise maximum and during diastole, the rest state of cardiac cycle the pressure is declines. The patient at normal resting state will have a systolic blood pressure of approximately 120 mm Hg and a diastolic blood pressure of approximately 80 mm Hg.

The blood that pumped out by the heart that reaches to our body circulation which can be divided into two types—systemic circulation and the pulmonary circulation. During systole of heart, the pressure in the systemic circuit is always high (120 mmHg) as the blood that leaves the heart and need to reach the whole body of the circulatory system. However, the pressure in major veins is as little as 4 mm Hg. But the blood from the right ventricle of heart that reaches into the pulmonary circulation which is relatively at low pressure as the function of the lungs is to exchange gases. Pulmonary circuit pressure has a systolic peak of 25 mm Hg and a diastolic value of 10 mm Hg. Therefore, blood pressure is pulsatile and dependent on the point of measurement. Clinically, the doctor is interested in the systolic and diastolic reading and the average pressure.

In a simple word, heart beat is nothing but a cardiac cycle which comprises one systole and one diastole. Heart rate is number of heart beats per minute, the heart rate is not directly related to the blood pressure as the impedance offered by the body is not constant. For example, a sudden scare happened when a piece of glass has broken on the morning in an urgent work. The person at this situation will have a physiological effect known as the 'fight or flight' response. Following a sudden excitement, the blood vessels near the skin's surface contract and vessels are constrict. Therefore, it reducing the blood flow and minimizing the loss of blood from wound. Blood is also diverted from circulation of other parts of body like stomach and intestines. In that way, it will reduce loss of blood in wounded area by avoiding blood flow that causes enlargement arteries and veins. This causes a change in composition of the blood which increase the clotting mechanism. Using the analogy of relating the blood pressure to voltage and the flow rate to current and the circulation system to impedance. The blood pressure is clearly dependent on the heart rate, the flow rate and the impedance characteristics of the circulation system.

9.2 SPHYGMOMANOMETER: BLOOD PRESSURE MEASUREMENT

The sphygmomanometer is the most common method for measuring blood pressure. As it is noninvasive procedure, it can be performed by your general practitioner. In general, blood pressure can be measured by two methods:

i. Direct method, and
ii. Indirect method

9.2.1 Direct Method

Direct method can be performed only in animals. Indirect methods are non-invasive methods can be used both animals and human beings. Direct methods can be used to measure the amount or volume of blood eject from the heart by using cardio-meter and flow meter. The main disadvantage of direct methods are,

i. Can be used only in animals
ii. Sometimes blood vessel can be cut to measure the flow which is risky to animal's life.
iii. In cardio-meter, the size of meter should be in the size of the heart but not always.

Several methods can be followed to measure the amount of blood ejected by the heart. The indirect method has some advantages and disadvantages.

9.2.2 Indirect Method

Brachial artery is usually selected for the measurement of BP because of expediency. The arm cuff of sphygmomanometer is tied around upper arm just 2 inch above the cubital fossa. Cuff should not be too tight or too loose and at the same time it should connect with sphygmomanometer. Now, blood pressure can be measured by three methods.

1. Palpatory method
2. Auscultatory method
3. Oscillatory method

In general, these methods are safe and preferred for measurement of BP. Sphygmomanometer—blood pressure measurement is one of the indirect methods to measure blood pressure.

To perform the measurement of blood pressure is carried out by placing the cuff around the patient's arm just 2 inches above the antecubital fossa. The wrapped cuff is inflatable and consists of a band with cloth sleeves and can be tied or strapped Velcro. This cuff is connected to a hand pump and a mercury manometer. When the cuff is inflated, the pressure in the cuff increases and the tissue in the patient's upper arm become compressed which occludes the flow of blood in the brachial artery as show in Fig. 9.1. The principle logic behind the measurement is when the pressure in the cuff is slightly higher than the pressure in the arteries of blood vessel, it occludes the flow of blood. The physician listens to the flow of blood at a point on the surface of brachial artery, i.e., just below the bag with a stethoscope. The pressure in the cuff is increased above the point where the blood flow occluded and stops in the artery.

In Fig. 9.1a, according to the principle, when the pressure in the cuff is greater than the pressure of artery of blood vessel there is no blood flow during the cardiac cycle. It is occluded.

In Fig. 9.1b, when the pressure in the cuff is greater than diastolic but less than systolic pressure during cardiac cycle a small of blood flow. Now the flow is called turbulent flow which is audible sound that can be heard in auscultatory procedure. Here the phases of heart sounds start.

In Fig. 9.1c, now the pressure in cuff pressure is still reduced but here the cuff pressure still greater than diastolic and less than systolic pressure. Therefore, there is a clear pitch of sound is heard in stethoscope where the heart sound will be in third phase. However, still there is noise associated with turbulence flow.

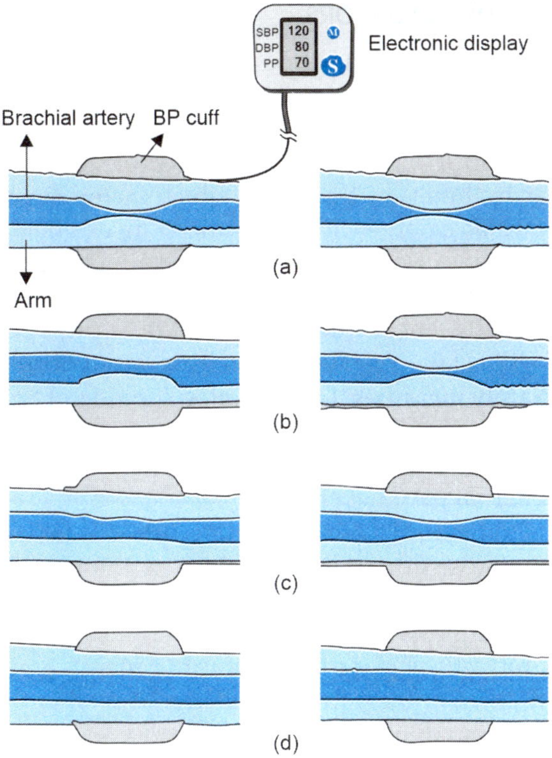

Fig. 9.1a to d: Principle involved in recording of blood pressure

In Fig. 9.1d, when the cuff pressure is less than systolic and diastolic blood pressure which means the pressure in cuff is now less than the brachial artery pressure now the vessel is fully opened. Turbulent flow of blood now become laminar flow. There is no audible sound which concludes the Korotkoff sound phases are finished.

During auscultation method of measuring BP, when the sound appears that time BP monitored with value as systolic blood pressure (normal range 110–140 mmHg), whereas when the sound disappears that time BP monitored as diastolic blood pressure (normal range 70–90 mmHg).

A number of different methods can be used to determine the point at which blood begins to flow again other than detecting the Korotkoff sounds. At beginning of this section, pressure in a cuff that cuts off the blood flow. As pressure is decreased the blood flow resumes when the cuff pressure is approximately equal to the systolic pressure. This flow may be detected by a Doppler ultrasound blood flow detector. The Doppler device produces an audible output which is proportional to the blood flow velocity. A phonocardiogram can be used in this principle only.

A transducer may be used to measure vibrations in the cuff which recommence as the cuff pressure approximately equals the systolic blood pressure. The vibration in the cuff reaches a maximum at a point when the pressure in the cuff is equal to the average blood pressure, and decreases and disappears when the cuff pressure is equal to the diastolic pressure.

Advantages: The sphygmomanometer is non-invasive method of measuring BP, easy to perform and the required instrumentation is also cheap. Measurements are with greater accuracy require invasive techniques.

9.3 DEFIBRILLATORS

The procedure in which an electrical device sends an electrical stimulus called shock that given to the heart to stop an arrhythmia. These results in the return of a productive heart rhythm and it is called defibrillation.

Defibrillation process is performed for the purpose to correct life-threatening arrhythmias of the heart, which includes ventricular fibrillation and cardiac arrest. In cardiac emergencies, this procedure can be performed immediately after identifying the patient experienced an arrhythmia that indicated by lack of pulse and unresponsiveness. For additional confirmation of arrhythmia, an electrocardiogram (ECG) can be displayed visually. For medical treatment by a physician, in non-life threatening situations, atrial defibrillation can be used to treat atrial fibrillation or flutter.

Safety measures: Defibrillation should not be performed on a patient who has a high level pulse or is alert, as this could cause a lethal heart rhythm disturbance or cardiac arrest. The paddles used in the procedure should not be placed on a woman's breasts or over an internal pacemaker.

In medical sciences, cardiac arrhythmias can cause irreversible damage due to prevent the heart from pumping blood to the body which includes the brain and heart. These includes ventricular tachycardia, fibrillation, and cardiac arrest. There are some chances for defibrillator to restart the heart that is fibrillated. Death can occur in minutes unless a productive heart rhythm, able to generate a pulse which is restored through defibrillation. Because immediate defibrillation is vital to the patient's survival.

To enable the heart's natural pacemaker to regain control of heart to beat, the defibrillators can deliver a electric shock to the heart. It is an electronic device that has a paddles and ECG monitor.

During this procedure, the paddles of defibrillator are placed on chest of patient with a suitable conducting gel that ensures good contact with the skin. During operating procedure of thoracic surgery, sterile internal paddles are applied directly to the heart. At this time, direct contact with the patient is discontinued by all caregivers. To increase the probability of further shocks for its effectiveness on patient in additional defibrillation, it should be repositioned. In the entire procedure, the patient's pulse and/or electrocardiogram is continually monitored when defibrillation is not in progress. Medications to treat possible causes of the abnormal heart rhythm may be administered. Defibrillation continues until the patient's condition stabilizes or the procedure is ordered to be discontinued.

Preparation: If a patient is found in cardiac anguish emergency, without a pulse and non-responsive, and help is beckoned, cardiopulmonary resuscitation (CPR) is begun and continued until paramedic services reach them and are able to provide defibrillation. Electrocardiogram leads should be attached to the patient chest and suitable gel or paste is applied to the defibrillator paddles and are placed on the patient's chest. The paramedic services people verify the patient with lack of a pulse while visualizing the electrocardiogram, assure contact with the patient is discontinued, and deliver the electrical charge.

Atrial defibrillation: "Atrial defibrillation is a treatment option that will be ordered for treatment of atrial fibrillation or flutter. The electrocardiogram will be monitored throughout the procedure. The paddles are placed on the patients chest with conducting gel to ensure good contact between the paddles and skin. If the heart can be visualized directly during thoracic surgery, the completely clean and sterile paddles will be applied directly to the heart. The defibrillator is programmed to recognize distinct components of the electrocardiogram and will only fire the electrical shock at the correct time. Again, all direct contacts with the patient are discontinued prior to defibrillation."

Pitfalls: The most common complications are observed in defibrillator is skin burns because the defibrillator paddles forms a electrical shock which may leads to skin burns in patients. Other risks include injury to the heart muscle, abnormal heart rhythms, and blood clots.

Normal results: This procedure can be performed to treat serious ventricular arrhythmia which is likely to be effective within five minutes of procedure and it can prevent brain injury and death by activating the heart to produce rhythmic normal pulse. After the emergency procedure, patients will be shifted to a hospital critical care unit for additional monitoring, diagnosis, and treatment of the arrhythmia. Intubation may be required for respiratory distress. Medications to improve cardiac function and prevent additional arrhythmias, are frequently administered.

Atrial defibrillation is successful at restoring cardiac output, alleviating shortness of breath, and decreasing the occurrence of clot formation in the atria. An AED, or automated external defibrillator, is used to help sufferers who are experiencing sudden cardiac arrest. It is a sophisticated medical device that easy-to-use and can analyze the heart's rhythm and if it is necessary to deliver an electrical shock, this defibrillation can perform better in emergency situations. The following protocol will be followed for AED to re-establish an effective rhythm of heart as shown in Fig. 9.2.

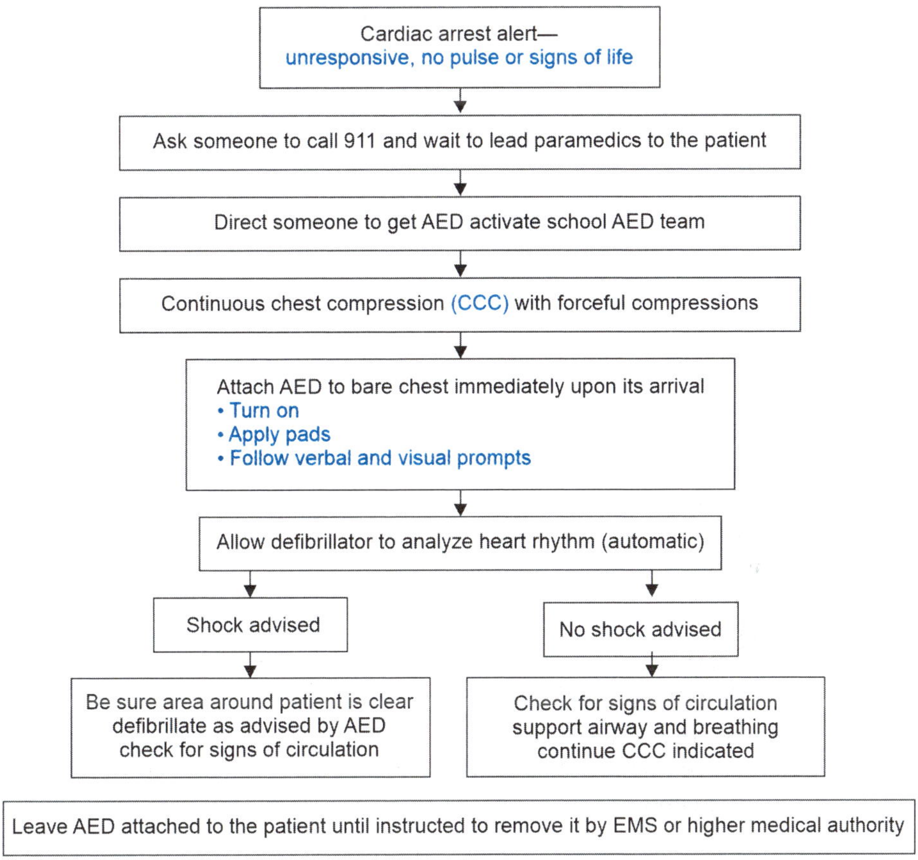

Fig. 9.2: AED protocol flowchart

9.4 CATHODE RAY OSCILLOSCOPE

The cathode ray oscilloscope (CRO) is an electrical instrument which is used for viewing waveform measurement and analysis part of waveforms. It has four major sections which are display, vertical controllers, horizontal controllers and triggers. It is showing graphical representation of waveform and depict like fast X-Y plotter that shows the input signal versus another signal or time. The CROs are used to analyze the waveforms, transient, phenomena, and other time-varying quantities from a very low-frequency range to the radiofrequencies.

Cathode ray oscilloscope working in voltage range and other physical quantities, like current, strain, acceleration, pressure, are converted into the voltage with the help of device called transducer. It is also used for time varying quantity like waveforms, frequency and transient phenomenon, etc.

The CRO has display area which shows a response to an input voltage as luminous spot. This luminous spot is produced with the help of electrons beam that striking on a fluorescent screen. The normal form of the CRO generated ramp voltage called "time base" which uses a horizontal input voltage. This luminous spot visualize accordingly as voltage moves of input periodically in a horizontal direction from left to right over the display area or screen. Whereas the vertical voltage part of waves moves up and down on the screen. The vertical voltage is the voltage under investigation. When the input voltage moves very fast on the screen, the display on the screen appears stationary (Fig. 9.3). Thus, CRO provides a means of the visualising time-varying voltage.

9.4.1 Construction of Cathode Ray Oscilloscope

CRO has some main parts which are as follows.

1. Cathode ray tube
2. Electronic gun assembly
3. Deflecting plate
4. Fluorescent screen for CRT
5. Glass envelop

Detailed explanation about the parts are given below.

1. **Cathode ray tube (CRT):** It is vacuum tube which converts the electrical signal into the visual signal with the help of electron gun and the electrostatic deflection plates (vertical and horizontal). Usually, the electron gun can produce a focused beam of electron which is accelerated to high frequency. The role of deflection plates can move the beams of electron up and down as well left and right accordingly. The vertical deflection plate moves the beams up and down and the horizontal beam moved the electrons beams left to right. These movements are independent to each other and hence the beam may be positioned anywhere on the screen.
2. **Electronic gun assembly:** The main role of gun assembly is to form a fine beam of electrons which are emitted from electron gun. The gun is already equipped with heater, cathode, a grid, a pre-accelerating anode, a focusing anode and an accelerating anode. To gain the higher emission of electrons, the layers of material like barium and strontium is coated on the cathode.

   Once the electron emitted from the cathode, it allows to pass through the control grid. It is mostly made up of nickel cylinder which is located co-axially. The main role is to control the intensity of emitted electron from cathode. After the emission of an electron from the cathode grid, it passes through the control grid. The control grid is usually a nickel cylinder which is located centrally with the CRT axis. The grid can control the intensity of the emitted electron from the cathode. Whenever, the electrons are passing through control grid, where it is accelerated by high positive potential of pre-accelerating or accelerating nodes.

The focused beam of electron is now subjected on focusing electrodes and then allowed to pass through the vertical and horizontal deflection plates. After coming out from the plates, the beams on placed onto the fluorescent lamp. The pre-accelerating and accelerating anodes are connected to 1500 V, and the focusing electrode is connected to 500 V. There are two methods followed to focus the electron beam. These methods are electrostatic focusing and electromagnetic focusing. The CRO commonly uses an electrostatic focusing tube.

3. **Deflecting plate:** Once the electron beam emitted from the electron gun that allowed to pass through two pairs of the deflecting plate namely vertical plate and horizontal plate l. The pair of plate is used to form vertical deflection and it is also called vertical deflecting plate or Y plates similarly, the pair of the plate which is used for horizontal deflection is called horizontal deflection plate or X plates.
4. **Fluorescent screen for CRT:** The face plate which in placed in the front of CRT. It has flat appearance in size up to $100 \text{ mm} \times 100 \text{ mm}$. But the screen of the CRT is slightly curved for larger displays. The face plate is formed by pressing the molten glass into a mould and then annealing it.

   Phosphor crystal can be used to coat the inside surface of the faceplate. The vital role of phosphor is to convert electrical energy into light energy. When a beam of electron strike on phosphor crystal, it raises their energy level and hence light is emitted during phosphorus crystallization. This phenomenon is called fluorescence.
5. **Glass envelope:** Glass envelope is completely evacuated and it is in the conical shape structure. The conducting materials like aquadag which is a water-based colloidal graphite coating is used in between the neck and the screen of inner surface of CRT. The coating material is act as a high voltage electrode. The coating surface is electrically connected to the accelerating anode and hence help the electron to be the focus.

Working of Cathode Ray Oscilloscope: Once the electrons are ejected from the electron gun, it allowed to pass through the control grid. The grid can control the intensity of electron in vacuum tube. If the grid has high negative potential, then it allows only a few electrons to pass through it because same negative charge repels each other. Thus, the dim spot is produced on the lightning screen. If the negative potential on the control grid is low, then the bright spot is produced. Hence the intensity of light depends on the negative potential of the control grid.

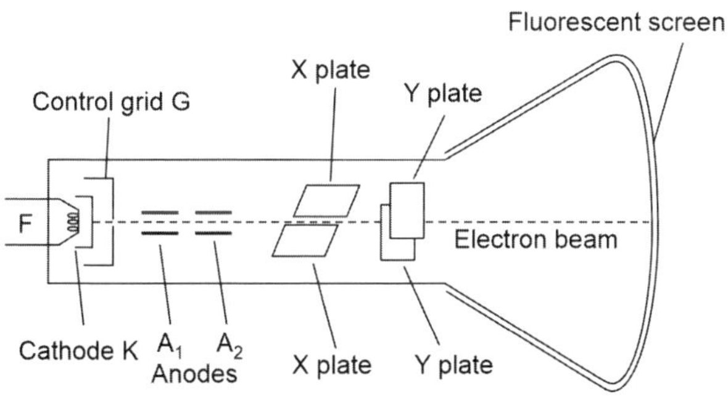

Fig. 9.3: Schematic diagram of cathode ray oscilloscope

After the role of control grid, the electron beam is now passing through the focusing and accelerating anodes. These anodes are at a high positive potential that can converge the beam at a point on the screen. After moving from the accelerating anode, the beam comes under the effect of the deflecting plates. If the deflecting plate is maintained at zero potential, the beam produces a spot at the centre. If the voltage is applied to the vertical deflecting plate, the electron beam focuses at the upward and when the voltage is applied horizontally, the spot of light will be deflected horizontally. In that way we can control the adjustment of CRO beams.

Applications

i. It can be used to measure ac and dc voltage supply.
ii. It can be used to study the wave nature of ac voltages.
iii. It can be used to find the frequency of ac voltage.
iv. It can be used to monitor the beating of heart in cardiology.

9.5 PLETHYSMOGRAPHY

It is an instrument that can be used to measure the volume changes in different rates and areas of the body. It has an ideal measure of changes within BP cuffs or other type of sensors. These are attached to a machines, called a plethysmograph.

It is particularly effective in detecting changes that caused by flow of blood in the body. It can help the doctor to determine blood clot in arm or leg. In addition, it can also help to calculate the volume of air in lungs that can hold.

Procedure for a plethysmography: If the person feels symptoms like redness in skin, warmth, swelling and tenderness due to blood clots in lower limb, the person is advised to take limb plethysmography. Commonly used methods to identify blood clots in limbs is an arteriogram. But in case of plethysmography, it is less invasive and economical procedure even though it has less accuracy when compared with arteriogram. Similarly, if the person feels symptoms of upper respiratory problems like pain or discomfort while breathing and shortness of breathing. The person is directed to have lung plethysmography.

9.5.1 Limb Plethysmography

This procedure can be performed after diagnosis of the patient who have sign of blood clots in limb. During the procedure the patient advised to wear hospital gown and asked them to keep one leg and one arm bare to perform test. The subject asked to recline in a comfortable position on an examination table.

Blood pressure cuffs will be placed on the patient's leg and arm (Fig. 9.4). The cuff pressure is increased by inflating with the help of hand pump of BP apparatus. The doctor will be concentrating on systolic blood pressure which is maximum pressure of the heart during contraction. The subject may feel a little uncomfortable when the blood pressure cuff is inflated and the test is performed about 20 to 30 minutes. During this procedure, the subject will be asked to move as little as possible. Limb plethysmography is not associated with any risks and side effects. Unless your doctor tells you otherwise, you can resume your regular schedule immediately after the test.

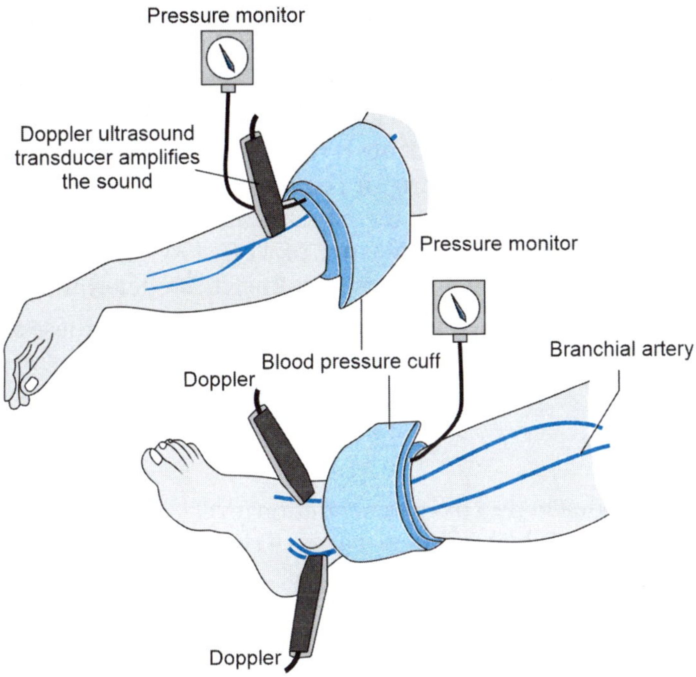

Fig. 9.4: Limb plethysmography

9.5.2 Lung Plethysmography

The procedure of lung plethysmography can be performed to the patient who have sign of respiratory problems like pain or discomfort while breathing and shortness of breathing. The subject is asked to sit in a small, airtight room and they will be clipped their nostrils to breath better against a mouthpiece. Sometimes, subject may report becoming short of breath or lightheaded. Let the doctor know if the subject experience these symptoms during the test. After the test and doctors advise, the subject can resume to their regular activities.

9.5.2.1 Understanding of Plethysmography Tests

Limb Plethysmography: In general, the systolic blood pressure in upper and lower limbs are similar. The ankle-brachial index (ABI) method is followed to check for potential problems that related to blood clot in lower and upper limb. This is a simple test that compares the BP in the upper and lower limbs. The ankle-brachial index test compares the blood pressure measured at your ankle with the blood pressure measured at your arm. A low ankle-brachial index number can indicate narrowing or blockage of the arteries in legs. A normal ABI falls between 0.90 and 1.30, beyond the range is blocked or narrow artery.

Lung Plethysmography: Lung plethysmography can measure the lung capacity to hold air inside. The normal range may vary from person to person and depends on age, gender, body size, and level of fitness. This is the initial diagnosis procedure, if any abnormal results that confirms the problem in the lung capacity. To confirm the severity level of abnormal lung capacity can be confirmed with additional test that recommend by the doctor. This abnormality may be due to a breakdown of lung tissue and problems with the muscles around chest wall. It may also include problems with lungs' ability to contract and expand. A lung plethysmography test can be done in many diagnostic labs like pulmonary function laboratories, cardiopulmonary laboratories clinics and pulmonology labs.

Total Lung Capacity (TLC-6l): The total amount of air that present in the lungs after a forceful inspiration is called TLC. It contains all the volumes of lung. If the volume of air present in the lung after a normal breathing out, which is called functional residual capacity (FRC), whereas amount of air is left when you breathe out as much as possible, or residue of air (RV) and the volume of air is residual volume.

As your chest moves while breathing, it changes the pressure and amount of air in the chamber. Breathing can also change the pressure against the mouthpiece. From these changes, doctor can get an accurate measure of TLC, FRC, and RV. While doing the procedure in the mouthpiece or against closed glottis, it can determine the values.

Risk factors: Risk factors associated with lung plethysmography, these are:

- Anxiety
- Dizziness
- Lightheadedness
- Shortness of breath

9.6 PULSE OXIMETRY

Pulse oximetry is technique to find oxygen saturation level in the blood. It is a noninvasive procedure and painless test which can detect a small change in efficacy of oxygen carrying capacity of blood that carried to the extremities which are furthest from the heart, including the legs and the arms. Sometimes, pulse oximetry is called the fifth vital sign, it measuring hemoglobin saturation by using a light signal that transmitted through the tissue.

It appears small, handy, and clip-like device that attaches to the place of collection of blood sample like fingertip, toes, or earlobe. Commonly, it is placed on a finger and often find in critical care unit, emergency rooms or hospitals.

9.6.1 Need to Know of Pulse Oximeter

Oxygen Saturation: Oxygen saturation is the proportion of oxygen that saturated to hemoglobin which is relative to its total hemoglobin in the blood. Pulse oximeters can measure the oxygen saturation level in the blood. Before learning to the principles of pulse oximeters work, we need to know about basics of oxygen saturation.

Oxygen plays a vital role in everyday life. Oxygen is mixed with atmosphere and it can enter the lung while breathing and then oxygen is transported into the blood. The major function of blood is carried oxygen into the various parts of our body. The main role is played by hemoglobin that carries oxygen in our blood. In general, oxyhemoglobin is nothing but the oxygen-loaded form of hemoglobin, the predominant protein in our red blood cells (RBCs). The hemoglobin without oxygen is called deoxygenated hemoglobin (deoxy Hb), whereas the hemoglobin with oxygen is called oxygenated hemoglobin (oxy Hb).

Oxygen saturation simply refers to the percentage of the availability of hemoglobin that carries oxygen. The following illustration shows the oxygen saturation clearly.

During the respiratory process, oxygen from the atmosphere can enter into the lungs through respiratory tract and to the respiratory membrane where transport of oxygen occurs. Now the oxygen from atmosphere reaches into the blood where it can bind into the hemoglobin of RBC. With the help of RBC, oxygen can travel throughout body and transport it wherever requires. In Fig. 9.5a, oxygen is binding with blood which is also called as oxygenated blood, whereas when the organs of our body needs oxygen to do work or even for metabolic process, the oxygen is dissociate from the blood and reach into the organ that require oxygen. Now the blood cell is called deoxygenated blood as shown in Fig. 9.5b.

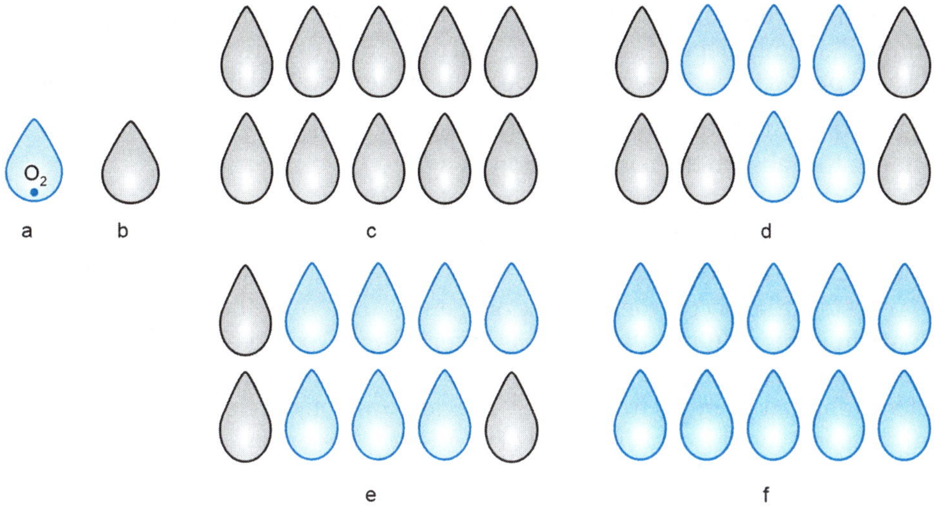

Fig. 9.5: Illustration of blood oxygen saturation

To understand better about oxygen saturation, let us consider the blood which has deoxygenated blood as shown in Fig. 9.5c. After respiratory process, hemoglobin of 5 RBC carries oxygen and percentage of the available hemoglobin that carries oxygen is 5 that mean oxygen saturation in ($5/10 \times 100\% = 50\%$). 50% oxygen saturation is illustrated in Fig. 9.5d, similarly 70% in Fig. 9.5e and 100% in Fig. 9.5f. So, oxygen saturation states about the percentage of the total hemoglobin that is carrying oxygen.

Physical properties: Oxygen saturation of blood can be determined by using infrared light that is emitted from light source. Once light emerges out form the source then it reaches into the light detector as shown in Fig. 9.6.

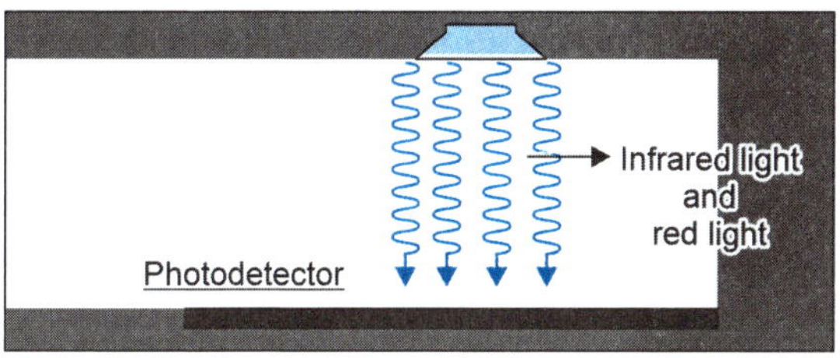

Fig. 9.6: Pulse oximeter light source

If a finger is placed in between the light source and the light detector, the light is now passing through the finger later, it reaches into the photodetector with some information. In general, part of the light is absorbed by the finger and the part not absorbed that reaches into the detector as shown in Fig. 9.7.

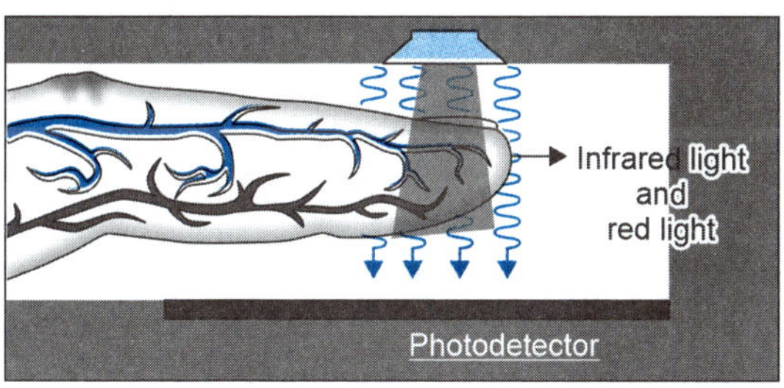

Fig. 9.7: Sensor orientation for light transmittance in the pulse oximeter

When the light from source is passing through the finger, the amount of light is absorbed by the finger is mainly depends on the physical properties of light that coming out from the finger. These properties are used by the pulse oximeter to calculate the oxygen saturation.

The amount of light absorbed is mainly depends on:
1. Substance that absorb the concentration of light.
2. Path length between light and absorbing substance
3. Oxygen concentration on blood that varies the absorption rate
4. Oxyhemoglobin and deoxyhemoglobin absorbs red and infrared light differently

Let us consider a subject to measure the oxygen saturation. We need to place the oximeter in the patient's finger. Once the finger is inserted into oximeter, the light from source which is present on the top of the finger emits light. Under the finger, we have detector that detect light that emerges from finger. The amount of light ejected out from the finger is mainly depends on the concentration of substance. For example, in the finger we have both artery which carries oxygenated blood and veins that carries deoxygenated blood. In the blood that contains hemoglobin (Hb) which absorbs light. The amount of light absorbed is proportional to the concentration of Hb in the blood vessel. If the blood vessel has a low Hb concentration that absorbs few amount of light and remaining maximum of light now reaches to the detector. The detector in the pulse oximeter can analyze the difference in incident and emergent light variations. This works under the principles of Beer-Lambert's law. According to this law, "amount of light absorbed is proportional to the concentration of the light absorbing substance".

By determining the amount of light that reaches the light detector, the pulse oximeter knows how much light has been absorbed. More the Hb in the finger, more is the light absorbed.

Depending upon the concentration as well oxygen concentration, the pulse oximeter can measure the oxygen saturation level in the subject.

Advantages of Pulse Oximeters
- Can be very compact and handy
- Cheap to buy and use
- Detects hypoxemia earlier than you using your eyes to see cyanosis.
- Non invasive

9.7 MEDICAL ULTRASOUND AND DOPPLER

Depending upon the ability of human can hear the sound, it is classified. Healthy young adult human can hear the sound in the range of 20 cycles per second or Hertz, abbreviated as Hz, to 20,000 Hz, or 20 kHz (kilo Hertz) termed audible sound (range of 20–20,000 Hz) (*see* Subsection 1.8.2.3). If the frequency of a sound is less than 20 Hz, it cannot be heard by humans and is defined as infrasonic or infrasound. If the frequency of sound is higher than 20 kHz, it cannot be heard by humans and is called ultrasonic or ultrasound. In general, typical frequencies used in medical ultrasound are 2–10 MHz (mega million, Hertz). Ultrasound frequencies that are commonly used in obstetrics and gynecology are between 3 and 10 MHz.

Generation of Ultrasound: The main source to generate ultrasound is a tiny piezoelectric crystal with transducers. When an AC is supplied to the crystal, it will contract and expand at the same frequency at which the current changes polarity and generate an ultrasound beam. The main advantage of ultrasound is the beam travels into the body at the same generated frequency. Once the high frequency ultrasound beam hits the internal organ and it will reflect and reaches into the transducer, now the crystals change it shape and this minor change in shape can generate a tiny electric current. The generated tiny current is amplified with suitable amplifier and sends the information to ultrasound machine to generate image on monitor. The Piezoelectric crystal with transducer works like transform electric energy into mechanical energy (ultrasound) and vice-versa. If the image is not clear because of insufficient ultrasound of one crystal for clinical diagnosis, the instrument is equipped with modern transducer with large number of crystals that arranged in parallel order that produce sufficient ultrasound beam as shown in Fig. 9.8. All crystals are supplied with electric supply and can be stimulated individually. The disruption of crystals damages are protected by rubber that helps to cover the crystal as well decrease the resistance from the crystal to the body. In practice, high frequency of ultrasound that produce by transducer cannot travel through air and will not reach the patient. To avoid this difficulties a suitable gel is applied on the patient that act as conductor and helps to permit the sound to pass in between patient and transducer. Modern transducers have crystals made of synthetic **Plumbium Zirconium Titanate (PZT)**.

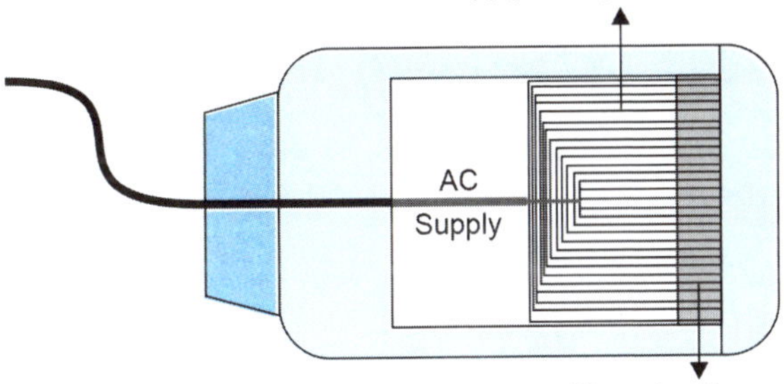

Fig. 9.8: Ultrasound transducer

Image formation of ultrasound: Modern ultrasound pieces of equipment are used to create a Doppler ultrasound image and it is sending multiple sound pulses from the transducer with different directions. It is analyzing the returning echoes of ultrasound that received in the crystals. A small physical properties can be used to understand the image formation in a better manner. For example, tissues of the body surface that are strong reflectors of the ultrasound beam, whereas bone or organ with hollow space produces a strong electric current. These currents are generated by the piezoelectric crystals which will appear as a hyperechoic (lots of echoes) image on the monitor. It can increase the density of sound waves when compared to surrounding structures.

Ultrasounds are impinged on human body or organ and reflected as echoes as a structure. In case of hyperechoic tissues generate a greater echo usually displaying as lighter colors during ultrasound imaging. On the other hand, weak reflectors of ultrasound beam, such as fluid or soft tissue, will result in a weak current, which will appear as a hypoechoic or anechoic image on the monitor. The ultrasound image is thus created from a sophisticated analysis of returning echoes in a grey scale format. Given that the ultrasound beam travels in a longitudinal format, in order to get the best possible image, keep the angle of incidence of the ultrasound beam perpendicular to the object of interest, as the angle of incidence is equal to the angle of reflection.

9.7.1 Uses of Ultrasound Scans

Abdominal scans: To investigate abdominal pain, nausea, vomiting, abnormal sounds and lumps, ultrasound scan is used. Organ structures like gallbladder, bile ducts, liver, pancreas, spleen, kidneys and large blood vessels can be observed. The demerit of ultrasound is hollow organ which means organ structures that contain air (such as the stomach and bowels) cannot be examined easily because air prevents the transfer of the sound waves.

Pelvic scans: Ultrasound can be performed to a woman who is suffering pelvic pain or has abnormal periods, fibroids, cysts or other conditions associated with the female reproductive system.

Pregnancy scans: Ultrasound can be used to check for fetal abnormalities, check the age and position of a fetus, and monitor fetal growth and development. Nowadays, undergoing an ultrasound scan is considered as routine checkup for pregnant women in India.

Other uses: Skeletal muscle scans (to check regions like a shoulder, hip or elbow), breast scans (abnormality in physical examination or mammogram) and an eye scan (internal structures) and to detect speed and direction of blood flow in certain regions of the body like arteries and veins.

Chapter

10

Electrocardiography

10.1 BASIC PRINCIPLE AND THE GALVANOMETER DEFLECTION

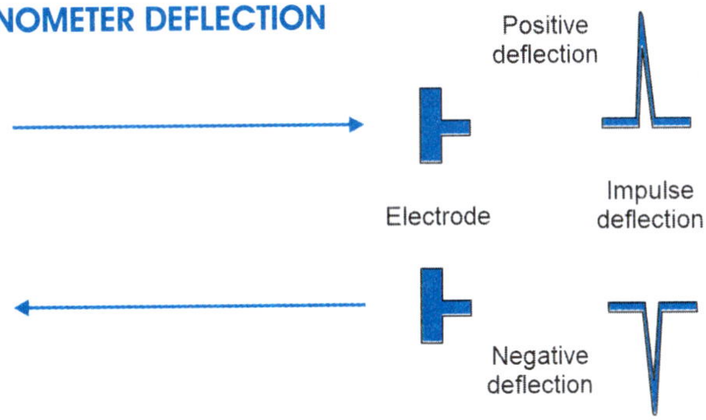

Fig. 10.1: Basics of electrocardiography

When an electric impulse flows towards a unipolar electrode or positive electrode of a bipolar lead, the galvanometer shows positive reflection or upward deflection. When an electric impulse flows away from a unipolar electrode or negative electrode of a bipolar lead, the galvanometer shows negative reflection or downward deflection and this is the basic principle of ECG and it is shown in Fig. 10.1. Flow of impulse in a cardiac muscle is occurs due to hemodynamic pump of the heart. For example, if atria contract blood flows from atria to ventricle in that way a small electrical impulse produces and this impulse of heart muscle conducts from atria to ventricle. If an electrode in placed at the bottom of ventricle during this sequence, the ECG curve shows positive reflection. Consequently, blood flows from ventricle to aorta that means flow away from the electrode and ECG curve shows negative deflection. This basic principle is involved in recording of ECG.

10.2 TWO CHAMBER CONCEPTS

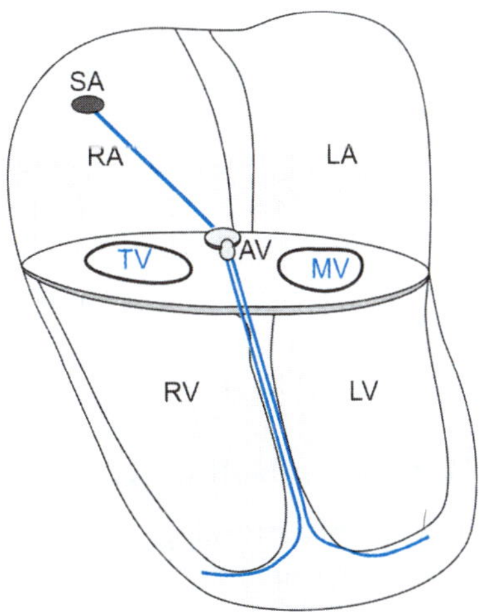

Fig. 10.2: Schematic diagram of heart chambers

In electrophysiology, it is stating that the heart is a four-chamber organ. In recording of ECG, when heart contract both the atrium contract simultaneously, whereas ventricle is relaxing on the same time. In the same manner both the atrium relaxing simultaneously, ventricle is in contracting period. So, in electrophysiological study, tee heart consists of only two chamber, i.e. one form atria and the other formed by the ventricles. In a simple understanding concept of electrophysiology, the two atria function as a single chamber of an electrophysiological unit, there is no electrical boundary between them and both atria are activated a single contraction or relaxation process.

This functional concept of chamber and its unit may be referred to as the biatrial chamber as well biventricular chamber. The two chambers atria and ventricles are separated from each other by an electrically inert conduction.

The fibrous A-V ring has this property and the possible communication between the chambers is possible only through the specialized conducting system of heart which is formed by AV node, the bundle of His and branches as shown in Fig. 10.2. Before going to the topic ECG, we have know about some terms and definitions.

10.3 ECG SPECIAL TERMS

10.3.1 Electrocardiography

A technique in ECG is called electrocardiography. In this technique is used to study about the electrical activities of the heart. The electrical impulses are spread and excited over through myocardium that produces electrical potential. Usually, this produces a mild current with low intensity and the current flows through the body that acts as a volume conductor. This tiny current can be recorded with a suitable electrode that may be place on the surface of the body and recorded in the graphical form is called electrocardiogram. This technique was discovered by Dutch physiologist, Willem Einthoven, who is considered the father of electrocardiogram (ECG).

10.3.2 Electrocardiograph

Electrocardiograph is the type of instrument or machine that works on the principle and technique of deflection (*see* above). Simply, it is instrument, that is, used to record and monitor the electrical activities of the heart.

10.3.3 Electrocardiogram (ECG)

Electrocardiogram (ECG or EKG from electrocardiogram in Dutch) is the graphical representation of the electrical activity of the heart. It is the sum up the electrical activity of all cardiac muscle fibers recorded from surface of the body.

10.4 TERMINOLOGY OF ECG DEFLECTION

The electrocardiographic deflections are named a letters instead of usual alphabet A, B, C, D, E and F, it is named arbitrarily and sequentially as P, QRS, T and U wave. Resting state of muscle which does not have impulse to conduct is called polarized state. Atrial activation is produced P wave reflection in ECG and the ventricular part activation also termed as ventricular depolarization which is produced QRS complex reflection. The T wave reflection is the ventricular recovery which is also termed as ventricular repolarization. The genesis of U wave is by Purkinje fibers. In a simple word, contraction heart muscle is called depolarization state and relaxation of heart muscle is called repolarization state.

10.5 ACTIVATION OF VENTRICLES

In an electrocardiological sense of studies, the ventricular muscle composed of three muscle masses, namely right ventricular mass, left ventricular mass and ventricular septum as shown in Fig. 10.3a.

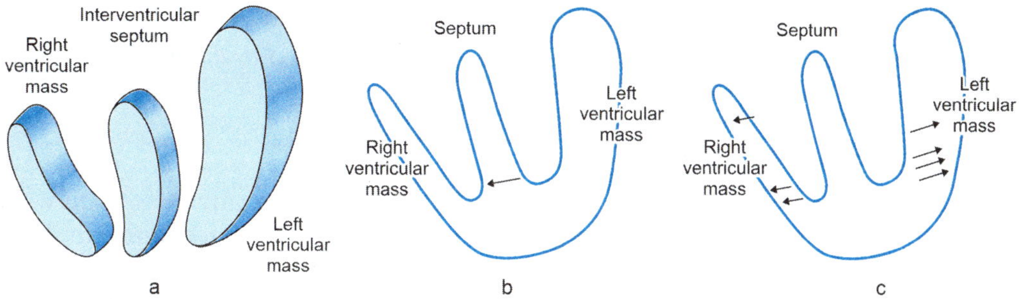

Fig. 10.3: Conduction in the heart

Conduction in the heart representing, activation is also called depolarization of the ventricles begins in the left side of the interventricular septum and spreads through the septum from left side to right side as shown in the arrow mark of Fig. 10.3b.

Naturally, left ventricular wall muscle is larger and hence a larger potential of electric force is generated when compared with right ventricular wall muscle and it is depicted in Fig. 10.3. Consequently, depolarization occurs both walls simultaneously and at the same time left ventricle is larger in muscle size that causes produce more electrical potential when compared with right ventricle and septa. This resulted in producing single force directed from right to left as shown in Fig. 10.3c. This is the main reason that we cannot observe the atrial repolarization curve in ECG.

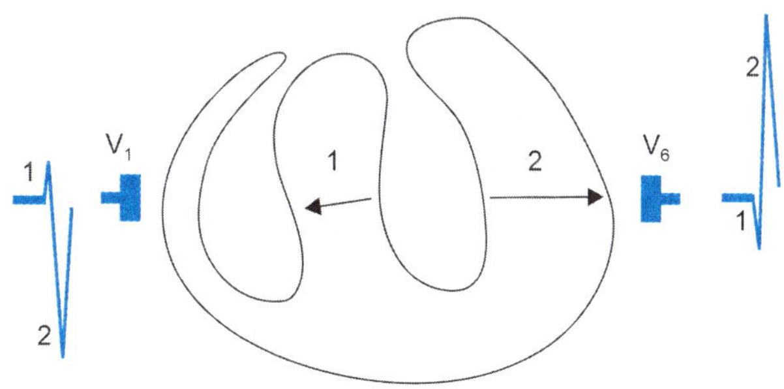

Fig. 10.4: Recording principle of depolarization and repolarization in ECG

Depolarization of ventricle form a small initial force from left to right through a septum that causes contraction of heart, followed by a larger force from right to left through the wall of left ventricle as shown in Fig. 10.4. These forces have magnitude and direction. In a physics, vector quantity has both magnitude and direction, and thus, it forms vectors. Vector cardiography is a separate chapter to study details form of ECG with vectors.

An electrode is placed to the left ventricle that is V_6 and it will record a small initial downward deflection caused by the stimulus spreading away from the electrode through septum which mention 1 in ECG curve of V_6. According to the basic principle, impulses are moving away from electrode V_6 that cause negative deflection followed by a larger upward deflection which caused by the spread of the stimulus towards the electrode through the left ventricular muscle. Thus, results in positive peaks in ECG as shown in point 2 of ECG in Fig. 10.4 of left ventricle side. Conversely, an electrode is place to the right ventricle of lead V_1 of ECG electrode will record a small initial upright deflection that caused by the spread of the stimulus towards the electrode through the interventricular septum followed with larger downward deflection that caused by the spread of the stimulus which is away from the electrode through the wall of left ventricle muscle. In the above cases, both the electrode records activity and impulse deflections of left ventricle alone that sequence order has been mention as 1 and 2 of the directions of impulse of the left ventricle. Similarly, all the leads from V_1 to V_6 records the ECG in the same manner but it records the summation of all the activities of the heart during contraction and relaxation process during cardiac cycle.

10.6 ECG LEADS AND ITS NOMENCLATURE

In electrocardiography leads are differentiated as positive or negative electrode which could be theoretically oriented in any relationship to the heart. By conventional methods, it can be classified and totally there are 12 lead placements to study about the electrical activity of the heart. These electrodes are placed on the surface of the body to record the ECG and called ECG leads. It can be fixed on the limbs or placed on the chest and can be classified limb leads and chest leads. In general, ECG recording is done by using ECG machine which has 12 electrode and classified into two categories: (I) Bipolar leads and (II) Unipolar leads

I. **Bipolar leads:** It is also called standard polar limb leads. These leads are obtained by connecting two limbs with two leads and form a connected lead. Here both the electrodes are active electrodes in which one is positive electrode and other is negative electrode. Standard limb leads are: LI, LII and LIII.

Einthoven, Dutch physiologist who was awarded the 1924 Nobel Prize for his discovery of the electrical properties of the heart through the electrocardiograph. By using the principle of Eienthoven, the electrodes are placed on the limbs. Usually, right arm, left arm and left foot are used to record ECG. When it is connected to the limb that form an imaginary equilateral triangle and the heart is considered at the center of triangle. This triangle is called **Einthoven triangle**.

Einthoven law: Einthoven's law states that the algebraic sum of the potentials of Lead I and Lead III is equal to that of Lead II and assuming that the heart will be at the center of the equilateral triangle. In simple words, the sum of the potentials of the all leads will be equal to zero. It is expressed mathematically as:

$$\text{Lead I} + \text{Lead III} = \text{Lead II}$$

or

$$E_1 + E_2 + E_3 = 0$$

Position of electrodes for standard limb leads is shown in Table 10.1.

Table 10.1: Position of electrodes for standard limb leads

Lead	Polarity	Lead Axis
Lead I	Right arm (–) to left arm (+)	± 180° – 0°
Lead II	Right arm (–) to left leg (+)	–120° – +60°
Lead III	Left arm (–) to left leg (+)	– 60° – +120°

II. **Unipolar leads:** In unipolar leads only one electrode is active electrode and the other electrode is indifferent electrode which is maintained at zero potential. Active electrode is positive and the indifferent electrode is serving as composite negative electrode. It is classified into two types:

1. Unipolar limb leads
2. Unipolar chest leads.

1. **Unipolar limb leads:** All unipolar leads consist of extremity of limb and precordial or chest leads. Extremity leads have a low electric potential, so these leads are

augmented instrumentally to increase the potential. These augmented leads are prefixed by the letter 'A'.

Unipolar limb leads are of three types:
(i) AV_R lead, (ii) AV_L lead and (iii) AV_F lead

AV_R lead is the augmented unipolar right arm lead as shown in Fig. 10.5. Electrically, active electrode is connected to right arm and indifferent electrodes which is formed by connecting other limbs together and with zero potential.

AV_L lead is the augmented unipolar left arm lead. Electrically, active electrode is connected to left arm and indifferent electrode is formed by connecting other limbs together and given zero potential.

AV_F lead is the augmented unipolar left leg lead. Electrically, active electrode is connected to left leg and indifferent electrode is formed by connecting other limbs together and given zero potential.

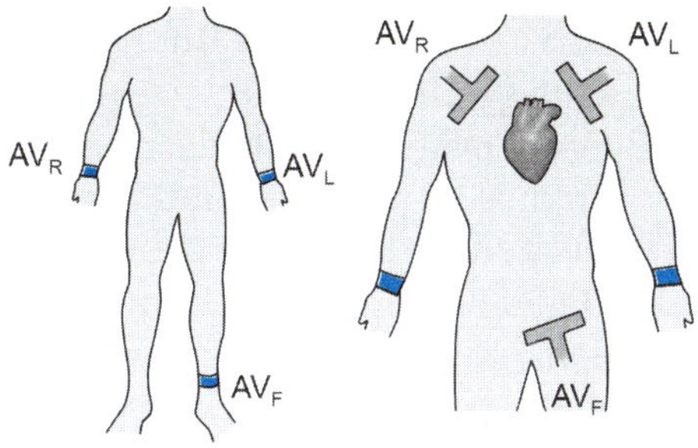

Fig. 10.5: Augmented unipolar limb lead

2. **Unipolar chest leads:** It is also called precordial leads which are designated by the letter 'V'.
   i. Lead V_1 is located over the fourth intercostal space to the right sternum border.
   ii. Lead V_2 is located over the fourth intercostal space to the left sternum border.
   iii. Lead V_3 is located equidistant between V_2 and V_4.
   iv. Lead V_4 is located on the midclavicular line over the fifth intercostal space.
   v. Lead V_5 is located on the anterior axillary line at the same level as V_4.
   vi. Lead V_6 is located over the midaxillary line at the same level as V_4 and V_5.

   The position of all 6 leads are shown in Fig. 10.6.

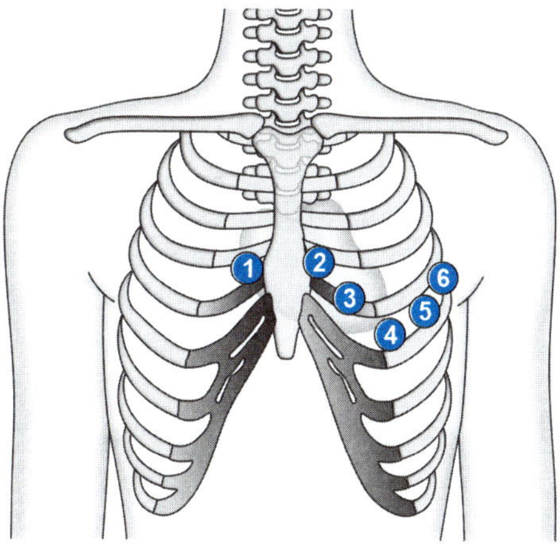

Fig. 10.6: Unipolar chest leads

Uses of ECG: Electrocardiogram is useful tool to diagnose the heart problems in routine and emergency situation.
1. Abnormal electrical conduction
2. Coronary artery disease
3. Heart attack
4. Heart rate
5. Heart rhythm
6. Hypertrophy of heart chambers
7. Poor blood flow to the heart muscle leads to ischemia

Bibliography

1. Avijit Lahiri. Basic Physics Principles and Concepts, fourth revision (an ebook), Lahiri, author and publisher, Kolkata, 2017.
2. Barry N Taylor. The International System of Units (SI), NIST Special Publication Edition, 2001.
3. Daenotes (internet), electronics–basic-electronics–atom-energy-levels-shells [place unknown: publisher unknown]; Available from: daenotes.com/electronics/basic-electronics/atom-energy-levels-shells.
4. Halliday D, Resnick R and Walker J. Fundamental of Physics. 10th Edition, Wiley and Sons, New York, 2014.
5. K Thayalan. The Physics of Radiology and Imaging, first edition, Jaypee Brothers Medical Publishers, 2014.
6. Kyle Kirkland. Light and Optics, an imprint of Infobase Publishing, New York, 2007.
7. LeoSchamroth. An Introduction to Electrocardiography, Blackwell Scientific Publications, 1976.
8. R Murugeshan. Electricity and Magnetism, 10th edition, S. Chand Publishing, New Delhi, 2019.
9. R Murugeshan and Kiruthiga Sivaprasath. Modern Physics, Seventeenth Revised Edition, S. Chand Publishing, New Delhi, 2014.
10. Sembuling K and Sembuling P. Essential of Medical Physiology. 6th Edition (New), Jaypee Brothers Medical Publishers, Delhi, India.
11. Shobhit Mahajan, S Rai Choudhury. Electricity, Magnetism and Electromagnetic Theory, Tata McGraw Hill Education Private Limited, New Delhi, 2012.
12. Uwe Krey, Anthony Owen. Basic Theoretical Physics—A Concise Overview, Springer Berlin Heidelberg, New York, 2007.
13. VK Mehta, Rohit Mehta. Principles of Electronics, Eleventh Edition, S Chand Publishing, New Delhi, 2022.

Index

U

V

W

Z